SUPERCHARGE YOUR BRAIN

SUPERCHARGE YOUR BRAIN

How to Maintain a Healthy Brain
Throughout Your Life

JAMES GOODWIN, PhD

PEGASUS BOOKS

NEW YORK LONDON

SUPERCHARGE YOUR BRAIN

Pegasus Books, Ltd.
148 West 37th Street, 13th Floor
New York, NY 10018

ISBN: 978-1-64313-867-1

10 9 8 7 6 5 4 3 2 1

Printed in the United States of America
Distributed by Simon & Schuster
www.pegasusbooks.com

Nature is the source of all true knowledge.
She has her own logic, her own laws, she has no effect
without cause nor invention without necessity.

Leonardo da Vinci

CONTENTS

" Dorothy gazed thoughtfully at the Scarecrow."

The Yellow Brick Road

If you only have brains on your head you would
be as good a man as any of them, and a better
man than some of them. Brains are the only
things worth having in this world, no matter
whether one is a crow or a man.

The Wonderful Wizard of Oz

IT WAS 1900 WHEN L. Frank Baum wrote *The Wonderful Wizard of Oz*,
and over the century and more since then some things haven't changed.
The Yellow Brick Road still captures an essence of the human experience.
Doesn't the search for an improved life or an improved self resonate with
us all? The Tin Man sought his heart along the road. The cowardly Lion,
his courage. For the Scarecrow – perhaps an unlikely hero for twenty-
first-century science – it was his brain.

It will be a long road for science, seeking an understanding of the brain
and unlocking its secrets; but already the hows, whys and whats of the
brain are unfolding on our path of discovery.

Arguably, this scientific journey started in 1848 in the US state of
Vermont, where a railroad accident overturned fifteen hundred years of
received wisdom about the brain. Phineas Gage was an affable and reliable
man; a respected foreman on the construction of the track between

Cavendish and Burlington. Distracted by an altercation between two of his men as he packed blasting powder into a granite hole, Gage allowed his concentration to lapse. The 6-foot tamping rod was blown through his head, entering under his left eye, exiting through his left cranium and landing some 30 feet away. Remarkably, Gage recovered. *But Gage was no longer Gage.* Once sober, industrious and temperate, now he was irascible, profane, unreliable, raging in fits of anger and abuse. Inconceivable as it seemed at the time, could the damage to the brain he had suffered be responsible for such a transformation in character, personality and temperament? Gage's physician Martyn Harlow, in his famous paper on 'The passage of an iron bar through the head', published in the *Journal of the Massachusetts Medical Society* that same year, concluded that it was.

The case of Phineas Gage and the revolution in thinking it created was a pivotal moment in our understanding of the brain. No longer could organic substrate (the 'grey matter') be dismissed, as the ancients had done, as 'cranial offal'. Over the following decades, as advances

Phineas Gage holding the tamping
iron that injured him

in empirical methods, imaging technology and hypothesis testing generated increasing numbers of scientific publications, the picture became clearer and we reached our modern view of the brain as an organ of immense complexity. With the false sciences of phrenology and mesmerism giving way to the real sciences of psychology, neuroscience and psychiatry, the brain ceased to be the 'ghost in the machine' – to the point where the axiom 'no consciousness in the absence of organic matter' is now an established tenet of brain science.

In 2000, one serendipitous event marked a huge leap forward. In July that year, two Scottish scientists were foraging about in the basement of the Moray House School of Education when they happened across a fifty-year-old treasure trove. It was the collected and archived results of one epic day's work in 1947, when Scotland had done something never attempted by any other country: it tested the IQ of over seventy thousand children. This was nearly the entire nation's population of 11-year-olds. And the unlikely reason for the investigation? It was thought that the working classes of Scotland were having too many children and were diluting the nation's IQ (they weren't). These unique data offered a glittering opportunity. Taking them as a starting point, researchers have been able to add a lifetime's worth of information about these people – their individual medical records, occupations, health concerns, lifestyles and environmental influences: in short, to map out their lives and see how their thinking skills have changed over time. This priceless opportunity was turned into a scientific study. And revealed a recipe for brain power.

Called 'The Disconnected Mind', this sparkling study is now in its twelfth year. It has given us information about many of the changes we can expect to occur on the journey of ageing: who stays sharp and who doesn't; what role exercise and activity play; the effects of alcohol, smoking and sex; whether friendships matter; and the impact of stress, poverty and 'socio-economic class'. This information is our path down the Yellow Brick Road, our journey to discovering the secrets of a healthy brain. Exploiting the participants' long lives, the study is now in its third wave, measuring and testing about a thousand of the Lothian Birth Cohort of 1936, individuals now in their eighties.

There have been some stunning findings. For example, in a paper published in *Nature* in 2012, the study settled one of the most controversial issues in brain science: is it nature (genetics) or nurture (environment) that determines our IQ? It turns out that 50 per cent of our adult intelligence can be accounted for by IQ in childhood (at age 11). But what of other factors? It turns out that only one-quarter of the change in our intelligence – our brain power – over our adult life is determined by our DNA. Fully three-quarters of that change is determined by our environment and lifestyle factors – in other words, *by what we do*.

What else is science revealing about the brain? The rate of progress in neuroscience and psychology is prodigious. A quick list of some of the biggest breakthroughs in just the past two years will show you why I say this:

- *in vivo* cell conversion, where gene therapy is used to reprogram the brain's support cells (glia) into working brain cells;

- the idea that the neurodegenerative diseases of the brain, many of which are incurable, are *preventable*, for example by dietary changes;

- the prospect of regenerative medicine being applied to the brain, by transplanting reprogrammed stem cells;

- the discovery of new links between exercise and cognition (thinking skills);

- improvement of memory by electric stimulation of the brain;

- opening the door to deliver therapies that cannot otherwise access the brain by using ultrasound and microbubbles;

- the discovery that frequency of sexual activity is related to cognitive improvement (I'm not joking).

However, there is a problem. It is that the messages arising from this new, revolutionary research are not reaching people in a balanced and well-informed way. There is hype, confusion, contradiction. Let me give you a prime example. In 2018, a study published in the *British Medical Journal* showed a greater risk of dementia in those who abstain from alcohol completely and those who consume more than fourteen units per week. In other words, moderate drinking confers a benefit. This finding was given great prominence in the media, including the BBC and other television news bulletins. Unfortunately, the message that moderate alcohol consumption reduces the risk of dementia was put out at a time when the prevailing advice from the Chief Medical Officer was that there is 'no safe level of alcohol drinking'. We can argue the nuances of the research and the rationale behind the messages, but that is not the point I'm making here – which is that, to everyone on the ground, it was terribly confusing.

Such confusions are compounded by the many books and publications which sensationalize, exaggerate or grind the axe of personal opinion about health-related issues. There has been a profusion of health-related books, promoted by aggressive marketing, which are grossly unreliable and misleading. Nowhere is this more obvious than in the world of food and diet – including their influence on the brain. Take, for example, the matter of dietary supplements. In the USA, there are currently 85,000 dietary supplements listed by the Food and Drug Administration, and in 2018 they generated sales revenue of more than $40 billion. The situation is much the same in the United Kingdom, and worldwide sales in 2018 were in excess of $121 billion. Millions of people are wedded to the idea that supplements are the answer to their health concerns. Like Dorothy and her companions, we can be seduced by the notion of a wizard who will wave a magic wand and deliver our desires. Why should we expect otherwise? With a little imaginative language, manufacturers can pretty much say what they like, as long as they don't claim to cure specified illnesses. An outstanding example is the sale of a jellyfish product that pronounces on the bottle: 'Improves Memory'. (It also pronounces itself vegetarian, which is interesting in itself since the active ingredient is an animal product – though in recent years it has been cloned.) Among the highly debatable claims the manufacturers make are: 'clinically shown to help with mild memory problems associated with aging' and 'supports a healthier brain, sharper mind & clearer thinking'. I believe none of these claims. And, as a founding member of the Global Council on Brain Health, I can tell you that they were met with derision when the Council reviewed them.[1] As I will reiterate many times in this book, I would urge you to be sceptical when you consider the many spurious and unjustified claims in the marketplace about what works and what doesn't work. Evidence is falsified, exaggerations are common and credibility is stretched – mostly in the pursuit of economic gain from an unsuspecting consumer. The key is to look very, very carefully at the evidence: both what it is and how it has been generated.

At the same time, there is a huge volume of technically complicated research findings being published about the brain. This literature is complicated because the brain is complicated – and the means by which it is studied are increasingly complicated too. It is voluminous simply because the pace of scientific advance in neuroscience, psychology and psychiatry is shockingly rapid. Day by day we are learning more. I once asked a very talented librarian what I thought was a quite straightforward question (it proved not to be so simple): 'How many scientific papers are published every day?' The answer, as near as we are likely to get, is three thousand. Three thousand per day. Since 1652, when the *Philosophical Transactions of the Royal Society* was first published, the rate at which these papers are published has doubled, on average, every nine years. The number itself is staggering. But more important is how many of them are relevant to the man and woman in the street, and of course how many of their benefits come to his and her attention.

One of the most profound revelations to emerge in the past few decades is that ageing is a lifelong process, that it begins early in life (around age 11), and that the rate of ageing in our bodies is not only malleable but largely under our control (DNA or heredity being responsible for only about 25 per cent of it). To a large extent, we can control our brain ageing, and with it our brain health, by modifying our exposure to risk factors such as diet, exercise, sleep, sex, alcohol, coffee, stress, social connections and how we use our brains.

Which brings us back to the Scarecrow and the Yellow Brick Road. As in *The Wonderful Wizard of Oz*, there is no magical wizard to give us what we want, no miraculous single solution. Instead, mere mortal scientists are working away behind the curtain, pulling levers, helping us to understand how to make the best of ourselves using the new, exciting and revolutionary findings about what is inside us.

Dorothy longs for 'someplace where there isn't any trouble'. When it comes to the brain, I'm not sure we've found that just yet – but it is not all obstacles and peril along our personal journeys. There are helping hands. This book, I hope, will be one of them. Inspired by my work with the Global Council on Brain Health, it will separate what is important from what is unimportant, what is myth from what is fact, what we know from what remains to be discovered. It will tell you what to expect, show you the landscape, and suggest how you can negotiate the likely twists and turns as your brain changes throughout your life. It will tell you how to avoid some pitfalls, and what are the real bullets you should dodge (such as being unfit, not getting enough sleep, being obese, being unsocial). It will help you to understand how to maximize, nurture and sustain your brain health; how to stay sharp, and how to beat the signs of decline as the years go by. It will advise you on strategies for staying vibrant and keeping brain health under your own control. It will empower you with the knowledge that you can get through life in one piece, minimizing loss of brain power or performance; that mental well-being gets *better* from middle age, and that the risks of ill-health, including dementia, can be much reduced. It will also offer you a look into the future, at how the trending science is revealing the secrets of improving brain power.

Meet your brain

IN 1953, TWO SCIENTISTS at the Cavendish Laboratory in Cambridge, James Watson and Francis Crick, discovered the genetic code. Consisting of a varied sequence of just sixty-four 'base pairs', it appeared deceptively simple. It took the best part of another half-century to arrive at the entire alphabet of human DNA – the genome. This was the achievement of the Human Genome Project, which began in 1990, ended in 2000 and cost $1 billion. It revealed that each one of us has a library of instructions for how to build a body comprising some 23,000 genes, made out of the 3 billion letters of genetic information in every cell. Now we know that this amounts to far more than just 23,000 single instructions. Each gene can code for tens of proteins – the basis of life in our cells.

If you think genetics is complex, consider the human brain: quite simply, the most complex structure known to science. So complex, indeed, that most neuroscientists will scoff at any notion that we understand it. Christof Koch, President of the Allen Institute for Brain Science in Seattle, admits: 'We don't even understand the brain of a worm.'[1] The worm he had in mind when making that comment, *Caenorhabditis elegans*, has a modest 302 brain cells and seven thousand connections – figures that rather pale into insignificance next to those of the human brain, which has 86 billion cells with thousands of connections *each*. And we are a long way from understanding even the numbers, types and function of all human brain cells.

However, what's important here is not what we don't know but what we do know. Since the concept of 'brain fitness' came into the mainstream in 2007, scientific findings about brain health have been prodigious in scale. Brain fitness is an idea based originally on the scientific finding of 'neuroplasticity' – the ability of the brain to form new connections throughout life, in response to trauma, disease, new demands (such as learning) and changes in the environment. I will come back to say more about brain plasticity later in this chapter. A decisive body of research findings has now shown that, as much as our genetics, our *lifestyle* is vital to the health of our brain. These new ideas are revolutionary. They show that brain ageing begins *early in life* and is connected to *how quickly we are ageing generally*. Equally revolutionary research has also shown that *we can put the brakes on these changes – and stay mentally young*. For example, brain health in our forties and fifties can be boosted *by engaging in specific activities in the twenties and thirties* – long before most people start thinking about brain health. Just the idea that we can slow down these changes would have been laughed out of court two or three decades ago. New research has, moreover, revealed that it is never too late to improve our brain health. Whatever decade of life we are in, we have more than a fighting chance of keeping ourselves sharp.

Our discovery of the brain and what it can do has been a long journey. We are going to start by looking at a revolutionary event some five hundred years ago.

A changing landscape

Summer 1543. Andreas Vesalius, soon to be appointed physician to the Holy Roman Emperor, had much on his mind. Unwittingly – and dangerously, in a time where unorthodoxy could easily cost you your life – he was about to revolutionize medicine by publishing a groundbreaking volume, *De humini corporus fabrica*, the first empirical atlas of human anatomy. The atlas reconciled two apparently contradictory forces: hard scientific empiricism on the one hand and artistic creativity on the other. Painstakingly researched by Vesalius and allegedly illustrated by a pupil of

Titian's, Jan Stephen van Calcar, it was a monumental work of genius. For the first time in the history of humanity, the drawings of the human body were not fanciful imaginations of the artist. Astonishingly for the time, they were based on dissections of human corpses – hitherto variously banned by the Church, by the civil authorities and by the received wisdom of Galenist medicine, which followed the Greek tradition in forbidding human dissection. As if by way of a prescient but ironic twist of fate, in 1514, the year Vesalius was born, Leonardo da Vinci had been forbidden by Pope Leo X to continue with his dissections, a prohibition that would later open the door for Vesalius' totemic volume. Vesalius' success was not simply the outcome of his talent, his novel approach to empiricism and his integration of art; it owed much to the fortuitous confluence of cultural change, European intellectual advances and new technology, principally the availability of Europe's greatest ever printer and block-cutter, Johannes Oporinus. Thanks to Vesalius, an accurate anatomy of the brain in all its natural splendour now lay exposed to the eyes of the world. But what the brain actually *did* was another question entirely. Paradoxically, though physicians were now privy to the stunning

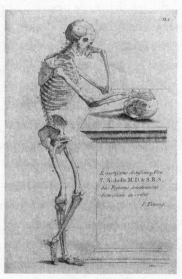

Meet your brain

beauty of the minutest intricacies of the brain, they understood nothing of its function.

Within decades of the publication of Vesalius' *magnum opus*, some hundreds of miles away in Stratford the Bard of Avon was wrestling with one of his mightiest works, *Henry IV, Part 1*. The four principal characters in Shakespeare's play were embodiments of a fallacious idea, embedded in human knowledge since ancient times, portraying human nature as a product of 'humours' in the body. Originating with the ancient Greeks, notably Aristotle and Hippocrates, this curious idea rested on the flawed perception that the body was possessed of four humours – black bile, yellow bile, blood and phlegm. The interaction of these humours explained the differences of age, gender, emotions and disposition. Their influence changed with the seasons and the time of day, and with the human lifespan. Heat stimulated action, cold depressed it. The young warrior's choler gave him courage; phlegm produced cowards. Youth was hot and moist, age cold and dry. Men as a sex were hotter and drier than women. The idea that the brain played any part in our natures would have been considered ridiculous – an absurd, risible concept.

The four humours

Such was the strength of these beliefs that Shakespeare was able to convey the commonly understood and accepted system of 'humours' in the four main characters of *Henry IV, Part 1* – even distributing the volume of text equally between them: Falstaff, phlegmatic; Prince Hal, sanguine; Hotspur, choleric; and the King, melancholic. Audiences loved the interplay of these characters and the underlying ideas – ideas that for over fifteen hundred years retained a grip on people, on their society, on science (such as it was) and on medicine. Until 1848. When an accident on a railroad in Vermont drove a tamping iron through them.

I have already told the story of Phineas Gage. This gruesome incident served, serendipitously, as the moment when science had to face an unpalatable truth. The received wisdom of medicine, stemming back over fifteen hundred years, was from beginning to end a fiction. For the first time, here was indisputable evidence that the seat of personality was the brain – not the heart, not the soul, and without doubt not the humours. Such incidents serve science well. Thomas S. Kuhn, a Harvard University professor, explained the utility of such moments in his 1962 book *The Structure of Scientific Revolutions*. In it, he argued that science, conservative in its approach, reserves its position until the evidence for a contrary view exceeds a critical mass, at which point the position changes. This Kuhn called a 'paradigm shift'. The revolution in thinking about the nature of the brain prompted by the case of Phineas Gage can fairly be deemed such a shift. The 1848 paper written by Gage's doctor, Martyn Harlow, did more than merely alert the medical profession to this gruesome case study – it set the ball rolling for a fundamental change in our understanding of brain function. Even today, two-thirds of all introductory psychology textbooks contain reference to the Gage case. And in enduring testimony to its importance, Gage's skull and tamping iron remain on permanent exhibition at the Harvard Medical School.

Our complex brain made simple

What we know about the brain can be summed up in a single aphorism. It is the most complex entity in the known universe. The latest methods in neuroscience have revealed that the adult human brain holds 86 billion neurons or nerve cells, each of which has 15,000 synapses or connections; 85 billion support cells called 'glia'; 528,000 miles of transmitting fibres; and a privileged blood flow of 750 millilitres (1.3 pints) per minute through 100,000 miles of vessels. In terms of cost to the body, the brain is an expensive organ. Its weight is just 2 per cent of the total, but it receives 15 per cent of the heart's output of blood. Its size has pushed the width of the female pelvis to its limits to achieve the birth of the large-brained newborn with the capacity for early walking, for language and for the social interaction necessary for group survival. And in terms of oxygen consumption, it is immensely thirsty, taking 20–25 per cent of the total for all metabolic needs, equating to some 500 calories per day – just for basic running costs. Why this startling level of metabolic privilege? Because the activity of the brain controls not only all aspects of all our basic survival mechanisms – body temperature, water level, acidity, blood pressure, hormone regulation, posture, balance, movement – but all our higher-level thinking (planning and decision-making), our social engagement with others and our emotional control.

In the most basic terms, however, the brain may be viewed as little more than a bag of salty water. Or, to be more precise, a bag consisting of a long, folded tube, with walls at their most only 5 millimetres thick. How could such a basic structure ever comprise the most elaborate organ in the body? We see the answer to this question in the development of the brain in the human embryo. Our brain begins to take shape at about twenty-one days after conception as a simple tube, the neural tube. From that point, at the front ('anteriorly') it greatly expands sideways ('laterally') either side of the mid-line. These expansions become the 'cerebral hemispheres' – the two halves of the brain. They are not completely separated but communicate with each other by means of a broad band of tissue, the corpus callosum ('hardened body'), and there is evidence that the female brain becomes more adept than the male at this 'cross-talk'. Researchers at the University of Pennsylvania scanned the brains of four hundred males and five hundred females aged between eight and 22, and after age

13 they found far more connections between right and left hemispheres in the female brain, facilitating more emotional processing and therefore social interaction.

Next comes a remarkable development. Starting about thirty-five days after conception, the front part of the tube migrates upwards and backwards, folding over itself so that the tip lies at the back of the hemispheres. It is during this phase of development that the elaborate folding occurs to give the brain its familiar 'walnut' appearance as a mass of 'gyri' (folds) and 'sulci' (grooves). This folding takes place for one purpose only – to pack as much brain material as possible into the confines of our cranium. The thin walls of the folding brain surface are what we call the 'cortex'; this, were it to be unfolded, has a surface area of 1.5–2 square metres – about the size of two pages of a large newspaper. Only in this way can our 86 billion neurons coexist within our 1.4 kilogram brain. It's probably worth mentioning here that brain size differs considerably between people at maturity, and women generally have smaller brains than men. But before all the men start to get smug, it's also worth mentioning that brain size is only weakly related to general intelligence. In 2012, a review of the evidence by Professor Richard Nisbett of the University of Michigan concluded that there was no significant difference between the general intelligence of men and women – echoing the findings of Jensen in his classic work some forty years earlier.

The building of the brain is a mammoth task on a tight timeframe – some 250,000 neurons are produced *per minute* throughout the course of pregnancy. The neurons are so closely packed that their dark, DNA-containing nuclei give a grey tint to the cortical brain matter – our proverbial 'grey matter'. And yet the mature brain is so sophisticated that a two-year-old's is only 80 per cent of its finished size – it does not finish developing until about age 25. As they are produced, new cells move to a predetermined area of the brain and then turn into – 'differentiate' into – the specialist cells of that area, eventually forming the various structures shown in figure 1.1. The maintenance of the brain is so important that at least 50 per cent of it – a much higher proportion in the highly active, sophisticated cortex – is made up of support cells called glia, which protect and support functioning neurons. There are several types of glia, including oligodendrocytes, which insulate the brain's wiring (white matter). This white matter – so called because of its fatty content – along with countless fibres, forms the interconnections between all the brain's 86 billion cells, known as the brain's connectome. The hemispheres are connected by transverse fibres, such as those of the corpus callosum. Association fibres link the different regions within the hemispheres, and projection fibres link the regions to the spinal cord.

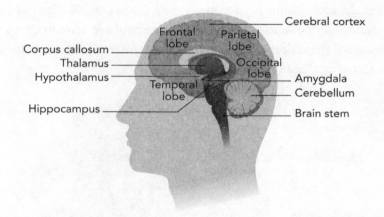

Figure 1.1: The human brain in situ

Situated above and below the corpus callosum are a number of bodies that together make up a powerful, ancient structure – the limbic system, known to psychology as the 'emotional brain'. Called 'limbic' from the Latin *limbus*, meaning border, this powerful system includes elements of our 'thinking' cortex – areas such as the hippocampus – and deeper, more primitive structures such as the hypothalamus, amygdala and thalamus, all shown in figure 1.2 overleaf. The hypothalamus is our critical control centre, regulating our hormones, sexual behaviour, blood pressure, temperature, hunger and thirst. The amygdala is our powerful 'anger machine', controlling anger, fear, anxiety and stress. Then we have the thalamus – a vast processing centre for all the sensory information coming into the brain from other parts of the body, such as visual images, sensations – including pain – and temperature. Critically, it tells the brain what is going on in and around our bodies – and reacts to all of it. It also plays a key role in our arousal or alertness levels. Not for nothing has the limbic system been called the 'feeling–reacting' brain as opposed to the 'thinking brain'. It has one, vital, overarching purpose: survival and self-preservation. Connected to the rest of the body by a two-way highway running in and out of the brain, it generates powerful reactions that permeate our whole being. Some behaviour we can't explain. Some we can't understand. Some is immensely difficult to control. The red mist of anger. The impulse to flee in terror. The compulsive drive to love, hate or enjoy. All these deeply felt experiences derive from the uncontrolled activities of our 'emotional brain'.

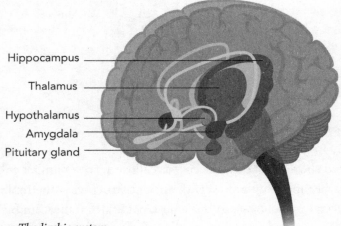

Hippocampus

Thalamus

Hypothalamus

Amygdala

Pituitary gland

Figure 1.2: The limbic system

The limbic system is dominated and kept 'under control' by the crushing weight of our social values, learned (or not!) and stored in the frontal lobes of the cortex. Temporarily incapacitate the frontal lobes through the anaesthetic consumption of alcohol, and uninhibited limbic behaviour frequently results – aggression, rage, anger, unbridled lust. Permanently disable the frontal lobes by a full-frontal lobotomy – a fate suffered by Jack Nicholson in the movie *One Flew Over the Cuckoo's Nest* – and pacification results, as the ability to scheme and plan is lost.

One of the emerging changes in thinking about the brain is a shift away from the idea of 'localization' of function – the idea, first put forward in 1878 by John Hughlings Jackson (a Fellow of the Royal Society and physician at the London Hospital), that certain parts of the brain are dedicated to a single function – movement, for example; or vision. The idea became embedded in medical orthodoxy after a furious debate in 1881 between the German scientist and vivisectionist Friedrich Goltz and David Ferrier, a pupil of Hughlings Jackson. Goltz lost that debate – but modern neuroscience has shown that he may well have been right, as it is becoming increasingly clear that the many areas of the brain work together seamlessly in a totally integrated manner. Surprisingly, for example, it's been found that more traffic occurs *from* the visual cortex *to* the thalamus than vice versa, even though we would expect the latter to be the case, given the routing role of the thalamus. It appears that as the visual cortex receives incoming messages from the eye, it constantly checks them against the resident images (our 'world-view'), accessed from the hippocampus via the thalamus, to understand what our senses are telling us. Sophisticated functioning requires a sophisticated, interconnected system.

And what of the unfolded remainder of the neural tube? It becomes the conduit of the brain–body tract: the spinal cord, the highway by which the most functionally advanced parts of the brain receive and convey life-sustaining and, for the most part, subconscious messages for which we give the brain little credit. With few exceptions, information about everything happening in the body finds its way into the brain through the spinal cord and cranial nerves, largely via our thalamus, that vastly powerful sub-cortical switchboard. Conversely, everything happening in the brain is directed by messages to the body. Without this vital reciprocal control pathway, everyday life would be impossible.

The lengthy and intricate development of the human brain makes us a species extraordinary, an exception to the evolutionary rules that apply to other primates. The frontal lobes represent a larger percentage of the cortex in the human (29 per cent) than in either the chimpanzee (17 per cent) or the rhesus monkey (11.5 per cent). But our higher thinking or cognitive skills are not attributable solely to our much larger 'pre-frontal' cortex. What matters is not so much size itself as the way in which the cortex is organized: the neurons of the human brain have a much more complex array of connections, which scientists call 'dendritic arborization' (a typically Latinate phrase meaning, essentially, a 'tree-like' pattern). And we have a much larger volume of white matter than could possibly be predicted from estimates of our grey matter. We truly have the most highly connected brain of any species – and that is particularly true of the female brain, which, as I mentioned a little earlier, shows more connections between left and right hemispheres – higher white matter complexity – than the male. It is no wonder that we became the supreme primate. In almost all respects, it would be true to say that whereas every other animal species on the planet survives by adapting to its environment, humans have survived as much by changing their environment as by adapting to it. By working together in ruthless social groups, and by working to temper primitive, emotional behaviour with the cold imperative of logic, we have harnessed the capacity of our large brains to overpower the natural world, inventing and reasoning ourselves to the top of the food chain. Little has been able to outpace the unstoppable momentum of this supremely intelligent species, empowered by group action, cooperation and the staggering capacity of the evolving human brain.

Brain health: the three core functions

Our ideas about health have changed dramatically in the past fifty years. No longer is it thought of as just the absence of disease. Now we see it as involving our capacity to cope with the changing physical, emotional and social pressures around us – our ability to adapt and self-manage. In no area is this more true than in the health of our brain – which, given the organ's vast complexity, could be conceived of as a maintenance nightmare. The crux of brain health is our ability to function well in daily life and work. Essentially, good brain health involves exercising three central functions of our brain: executive function (decision-making, problem-solving, reasoning, learning and memory); interacting successfully with others – what neuroscientists call 'social cognition'; and enjoying emotional balance or well-being.

Which television stations do you watch? What coffee did you choose this morning? Did you cook supper last night or eat out? Did you decide to walk the dog or just stay in? It sounds like a pretty truncated and mundane list of decisions, but neuroscientists at Yale have worked out that the brain (that is, every one of us) makes an astonishing 35,000 decisions *every day*. Assuming that we spend around seven hours asleep (blissfully decision-free), that works out at roughly two thousand decisions per hour or one decision every two seconds. To make these decisions, we need awesome computing power. And awesome it is: the brain has a memory capable of holding 1 trillion bytes of data and can turn over 100 trillion operations per second. Only the largest computers get anywhere close (more on this in a later chapter). This, in summary, is the executive power of the brain. Problem-solving, reasoning, learning all depend on it. This power has evolved from the need to out-think and out-perform the competition – whether predator, prey or a fellow human. In modern life we are not under quite the same survival pressures that our ancestors faced, but pressures we have none the less. Now new science is showing us how to stay sharp throughout our longer lives, how to resist the erosive effects of stress, and how to maintain our attention and think clearly in an age of information overload. Throughout this book, these pressures will be analysed chapter by chapter, along with strategies for handling them, to give you a bible of advice for maintaining your brain power. For example, it's now known that we can speed up the brain's activity through practice. In an experiment at Vanderbilt University, students had their multi-tasking tested. They had to identify which one of two possible faces appeared on a screen while responding to one of two possible sounds. After just two weeks of practice, the participants could do both tasks in rapid succession almost as quickly as doing either one on its own.

So much for executive function. What about social cognition? Social cognition is how the brain processes, stores and applies information about other people and social situations. This was at the root of what gave humans the edge over other primates. Around a hundred thousand years ago, in the competition for survival and dominance, the human brain developed superior cognitive networks for group working with other humans. This essential survival asset underpins all human society, all its rich social networks and the benefits they confer. And if humans got the edge over other primates, women got the edge over men. New science has confirmed what we always suspected: women are better at sending and receiving non-verbal social messages, including reading emotion in facial expressions, gestures and body language. Research shows that seventeen out of twenty women are more accurate at decoding social cues than the *average* man of the same age. These advantages are especially marked in female-only groups – that is, women read other women even better than they read men. These findings can be explained by the different ways in which the female brain is wired; and this difference in turn may be explained by the different selection pressures in biological evolution. In later chapters, we will see the devastating effect of social disconnection on how our brains function.

Reading other people's emotions is one thing. Managing your own is another. Keeping an even keel is a big, big deal. Without that ability, performance suffers and social relationships go out of the window. It is a high-stakes game. An errant individual in a hunting party could cost the group its prey – and with it, its survival. An emotionally out-of-control individual has the potential to wreck social and interpersonal relationships, costing the group its cohesion and again threatening its survival. The evolutionary key to this problem was the development of a huge pre-frontal cortex – our frontal lobes – providing massive inhibitory power to regulate the emotional drives of our so-called 'reptilian brain': anger, rage, aggression. Emotional balance occurs when we allow our conscious mind to understand, regulate and accept our feelings without letting them drive our behaviour. New findings are showing us just how and why these mood states arise – and what we can do about them. For example, in the past ten years the new science of psychobiotics has revealed that one big and surprising factor is the activity of the microbes in our colon, which can determine our emotions and our feelings – an amazing concept. We shall come back to this in chapter 4, 'Bugs in the brain'.

How our brain changes with time

Science can now show us what to expect of our brains as we travel through our lifespan. It's now possible to track our cognitive abilities or 'thinking skills' through the use of neuropsychological tests, and neuroscientists have found that most people are already showing declining 'spatial' skills by the mid-twenties. Our memory and reasoning start to slow down marginally in our early thirties, and by the middle of that decade we are processing information more slowly. Although these changes are initially subtle, they add up with each passing year and become more noticeable after age 50, when our 'forgetting moments' often start. In a study that examined the 'pace of ageing' in young adults aged 26–38, those who showed more general biological ageing – identified by a number of health indicators – than their (theoretically estimated) predicted rate also showed greater cognitive decline and brain ageing by as young as 38. They even looked older! In other words, the rate of brain ageing is related to the ageing of other bodily systems, such as the lungs and heart, the liver and kidneys, and the immune system, even at a young age. So it seems that we can predict how our brains are doing by looking at our general health. I'll return to this point later in the chapter.

We should all be concerned about maintaining our memory. It's as important as our physical health, if not more important. Your short-term memory enables you to remember the words you are reading so you can understand this sentence when you reach the end of it! Your long-term memories form the basis of your personality and your life. Memory, however, is much misunderstood. Evolution never intended us to have a video-recorded memory which would store everything we saw, heard or experienced. Our memory evolved to enable us to think, plan and do what we needed to do to survive, not to give us total recall. Our memory isn't even reliable. Far from it. For our 2 million years or so of human existence, it's worked well enough – as the expression goes, it's 'close enough for government work' – but it was never intended to cope with today's technologically laden world. And finally, there's no logic to the way our memories are stored, or which details we remember. We construct them on the largely emotional basis of what rewards us or what is important to us.

Take one feature of memory that drives most of us mad – the inability to put a name to a face. First, we should recognize that 'face blindness', as it's sometimes called, is experienced by everyone to a greater or lesser extent. It's just that younger people (in general) don't worry about it so much, whereas the rest of us become preoccupied with it and embarrassed by it. What neuroscience tells us is therefore very reassuring: we're bad at remembering names because there's no reason why we should be good at it. Faces carry meaning, tell a story, evoke our emotions. But names alone tell us very little. And further, our short-term memory will drop any meaningless item unless it is rehearsed. The lesson here is not to identify a common and apparent failing as a sign that we are 'losing it'. The evidence, to the contrary, is that we can stay sharp much longer than popular culture would have us believe.

Brain plasticity, also called neuroplasticity or neural plasticity, is the brain's ability to undergo structural and physiological change to cope with new threats and challenges. It enables us to respond and adapt as the result of experience. It occurs through changes in the synapses and the growth of new brain cells. Neuroplasticity was once thought to occur only during childhood, but research in the latter half of the twentieth century showed that many aspects of the brain can be altered (are 'plastic') throughout adulthood – that plasticity lasts, as Norman Doidge put it in his book *The Brain that Changes Itself*, 'from the cradle to the grave'. This means that many of us will be able to remain mentally acute and competitive, picking up new skills, throughout the greater part of our lives – even out-performing people much younger than ourselves as we deepen our knowledge and master new activities. Which is great news for anyone in fear of competition from younger, more ambitious and apparently 'sharper' colleagues.

There is no better example of brain plasticity than a hemispherectomy, a surgical procedure where one half of the brain is removed. This operation sounds too radical ever to consider, much less perform, so readers may be relieved to know that it is only carried out when the alternative is worse (for example, when death is threatened from a fatal disease, such as Rasmussen's encephalitis, in one half of the brain). Unbelievable as it sounds, the brain is so resilient and adaptable that even this extreme surgery has no apparent effect on personality or memory – they are conserved in the non-affected part of the brain, entirely due to the plasticity of the neurons. Certain brain functions which are unilaterally located (such as the speech centre in the left hemisphere) also appear to be taken up by the opposite hemisphere over time. There are some deficits that persist after surgery: for example, in a study of fifty-eight children who underwent this operation to control unremitting epilepsy, almost all emerged with 'hemiparesis' (one-sided weakness) in the contralateral arm (the one on the opposite side to the part of the brain removed), but all of them could walk and some could run. The point is that the anticipated effects of such drastic surgery are mitigated by the plasticity of the brain – in this case, its capacity to change and adapt after trauma. Plasticity is an inherent property of our brains and is the key to maintaining our brain functions.

Looking after our brains

How should we look after our brains? It is a daunting question. Longitudinal studies – where individuals are recruited as children or infants and then followed through their lives – are probably the best method of all for identifying the factors at play in keeping our brains sharp. Such studies can help to tease out the relative influences of genetics, epigenetics (changes due to how existing genes work) and environment. The evidence in the Disconnected Mind study has shown, for example, that 50 per cent of our intelligence in adulthood can be accounted for by our childhood IQ. Of the remaining 50 per cent, about one-quarter is attributable to our genes. That means that three-quarters of the change in our intelligence over the course of our lives is attributable to lifestyle: what we call 'modifiable risk factors'. We still have a long way to go in gathering enough evidence to be certain about the effects of each of these factors on our brain. The various chapters in this book deal with most of the important ones, such as exercise, sleep, sexual activity, social relationships, stress and well-being, and our gut microbes, nutrition and brain activities. However, probably the best general principle is to look after our general health – and, in particular, to keep on top of something we call low-level, long-term inflammation.

We normally think of inflammation as the acute response of the immune system to an injury or infection. The manifestations first described in AD 100 by the Roman author Cornelius Celsus as *rubor et tumor cum calore et dolore* (redness and swelling with heat and pain) became the hallmark clinical diagnosis of inflammatory disease. Essentially, inflammation is a protective reaction of the immune system. However, inflammation may also be chronic: that is, a low-level response over long periods of time to stress and to sustained long-term conditions, such as diabetes, obesity and arterial disease. It is a subtle but invisible process that is going on in our bodies on a daily basis and increases as we age; that is to say, the general level of inflammation in our bodies increases year by year. Even if we have no long-term illness and even if we are apparently healthy, a certain level of inflammation in the tissues will accumulate as our immune system responds to the challenges and stresses of daily life. Low-level inflammation in this respect receives little attention. Science has now identified 'inflammatory markers' – molecules which appear in the blood as a reaction to trauma or stress: cholesterol, C-reactive protein (CRP), cytokines, fibrinogen and many more, all of which are indicators or predictors of current, imminent or future illness. The immune system produces these molecules as a response to damage, infection or other changes in our tissues. For example, CRP acts as a 'surveillance' molecule, providing a 'look-out' or early warning signal for the immune system that something is amiss. But they also predict the rate at which our body and our brain are ageing.

This age-related increase in inflammation is an inevitable process; but the great news is that we now know that we can slow it down – and keep our brains more vital and healthy as we do so. If we reduce high levels of inflammation we slow down the overall rate of ageing in the body, reducing our risk of long-term illness, including brain conditions. On average, someone aged 30 or 40 will have no chronic illnesses; but by 65 the average person will have been diagnosed with one long-term illness, and the average 85-year-old will have five or six diagnoses and be taking ten to fifteen prescribed medications. Simple observation tells us that some people go on looking young, stay young in outlook and live long, engaged, productive lives. Sadly, for others, the reverse happens: early chronic illness, poor productivity, loss of independence and probably a shorter lifespan. Until the past few decades, these individual differences could not be adequately explained. Now we know that ageing is a lifelong process, beginning around age 11, and is subject to the influence of 'modifiable risk factors' throughout our lifespan; and that these factors are all, to some degree, inflammatory. I have mentioned some of them: diet and nutrition, sleep, social life, physical activity, alcohol, tobacco, drugs and stress. For example, every time we eat, the level of inflammation in the body rises, simply because we are ingesting foreign matter – and the more we eat, the greater the inflammation. Sleep badly and inflammation rises. Get 'stressed out' and our inflammatory markers rise. These changes can occur at any age. They apply at age 20, 30, 40, 50 and on until death. But as we get on in years, our bodies deal with them less well; so it is imperative to take action to reduce inflammation as we grow older. The relationship between age and inflammation is so strong that some scientists refer to the combination as 'inflammaging'.

It is true that as we become older, changes occur in all parts of the body, including the brain. Certain parts of the brain shrink, especially those important to learning and other complex mental activities, and communication between neurons can be reduced. Blood flow in the brain may also decrease, especially if the arteries harden. And we know that neuro-inflammation in the brain, like inflammation in the rest of the body, increases with age. But the extent of these changes can be slowed down – and that is the principal message of this book. Every diagnosis of dementia or 'mild cognitive impairment' (a decline in thinking skills) is preceded by twenty to thirty years of neurodegeneration and inflammation. Over 42,000 people under the age of 65 in the UK have dementia. So, in these people – astonishing as it may seem – the disease process would have started as early as age 35.

Dementia and other forms of cognitive decline are among the most feared aspects of later life – to the extent that they are often referred to as 'the new cancer'. In the 1950s, cancer was most often diagnosed late and couldn't be cured, leaving sufferers and those close to them with a sense of defeat and helplessness. Today, while cancer treatment has improved, our understanding of dementia and our ability to treat it are still very new science. Until recently, no one knew whether or not anything could be done to prevent the effects of degenerative change in the brain. Until the FINGER Study, that is.

On a cold December day in 2013, in Lancaster House, London, the then Prime Minister David Cameron brought together three hundred of the world's brightest brains to consider strategies to deal with Alzheimer's disease and dementia. It was an epic first, and Britain rightly received many accolades for convening this G8 conference on one of the most intractable and, some would say, unavoidable crises of our times – an inevitable result of our ageing population. Though the focus was on a cure (with a target date of 2025), in an unusual move the G8 mooted the idea that if we couldn't yet cure dementia, we might at least reduce the risk of it. Prevention was the dark horse coming in on the blind side. Few people were convinced. However, we already knew in 2013 that if you are middle-aged, of low educational attainment and have high blood pressure, obesity and/or diabetes; if you are physically inactive, a smoker or have depression – then there is a big chance you will eventually have dementia. The trouble was, the evidence was only by association. We didn't know for sure that if you removed these behaviours, then dementia would be prevented. But there was a lot of expectation riding on a study that was under way at the time of that conference; and a year and a half later, in June 2015, the findings of this study – the FINGER study – were published in the *Lancet*.

FINGER – the Finnish Geriatric Intervention Study to Prevent Cognitive Impairment and Disability – was undertaken over two years. It followed 2,554 people aged between 60 and 77, looking at the prevention of 'cognitive [brain] decline' by comparing two groups – one which followed advice on diet, exercise, cognitive training and vascular risks, and one which was given just general health advice. The result was clear, and startling: the interventions improved or maintained the cognitive status of at-risk individuals and reduced their risk of cognitive decline. The authors of the study concluded that if the beneficial effects on cognition observed in FINGER led to even a modest delay in the onset of dementia and Alzheimer's disease, they would have a huge effect at both individual and population levels.

So important are the results of this study that the World Health Organization has set up WW-FINGERS, a network of prevention (or, more properly, risk reduction) trials around the world. The 'take-home' message for us is that it's possible to reduce the risk of cognitive decline using lifestyle interventions. Quite simply, the study has shown that there is great value in reducing the risk factors in our lives, promoting our overall health and reducing our risk of getting chronic diseases.

And we can start doing this at any time, at any age. In the introduction, I mentioned the world-famous Scottish study 'The Disconnected Mind'. I was once asked on the BBC *Today* programme what was the most startling finding to emerge from it. For me, there was only one answer. In the original IQ test of over seventy thousand children, no one achieved a perfect score. In the same test over six decades later, many of the same individuals, who were now on average 74, did – to the astonishment of the research team. What a brilliant and incontrovertible response to all the cynics who say that we can't stay sharp, can't keep up, can't stay on the edge as we pass through 30, 40, 50 and beyond. History is full of people of all ages who have become great thinkers, leaders and achievers: Jo Pavey, who at 40 won the European gold medal in the 10,000 metres; Diana Nyad, who swam from Florida to Cuba at 64; Winston Churchill, who became Britain's wartime leader at 65; Lord Palmerston, Prime Minister at 71; Peter Roget FRS, who invented the thesaurus at 73; Nelson Mandela, who became South Africa's President at 76; and Dorothy Hirsch, who reached the North Pole at 89. The list is endless. And inspiring.

So: contrary to common belief, our brain performance won't necessarily go on a downward slope as we get older. The evidence shows clearly that some people *improve* mentally with age. This includes maintaining high childhood levels of intelligence – which are in most people very stable; but other mental skills, such as vocabulary and word use, *improve in almost everyone*. Psychologists refer to these skills as 'crystallized intelligence', and distinguish them from 'fluid intelligence', a more difficult skill to maintain. This 'fluid intelligence' is the general ability to think abstractly, reason quickly, discern relationships, recognize patterns and *solve problems*. Fluid intelligence can be your secret weapon. It's not dependent on education or qualifications. Nor on your experience or how much you know. But it will help you to be a great problem-solver – a very attractive proposition. It's the kind of skill you need for what is sometimes called 'out-of-the-box thinking', and it will make you innovative, creative and exceptional. How can you nurture it? That question has been plagued by forty years of uncertainty; but new evidence shows that fluid intelligence is to an extent trainable. By challenging the brain, working out of your comfort zone and taking on activities you normally shy away from, the chances are you will improve your fluid intelligence. One of the big messages of this book is: 'challenge your brain'. New experiences, new skills and even new threats are key to keeping your mind sharp.

Brave new world

Neuroscience, and the influence it has on our everyday lives, is advancing like a wildfire. Should anyone doubt this, I list here *just a few* of the many developments, many of them stunning and some even frightening, which have happened in the past two years alone. Take a look:

- It's now become possible to interface the brain with computers – an astonishing breakthrough with huge promise (and risks). An exciting study at the University of California, San Francisco, has shown that a computer may be capable of translating brain signals into words. And by late 2020, brain–machine interfaces were being developed by some US corporations, including among others Neuralink (founded by Elon Musk).

- A study in April 2019 found that the adult human brain is able to produce new neurons until the tenth decade of life – that is, our brain renews itself even into our nineties: clear proof of neuroplasticity and the potential for maintaining brain health and performance.

- Foods high in sugar have been found to reward the brain exactly in the same way as hard drugs, such as cocaine and heroin.

- Researchers at Johns Hopkins University have found that Parkinson's disease begins in the gut and migrates to the brain; and the Flemish Gut Flora Project showed that gut flora are related to depression. (I'll come back to this in more detail in chapter 4, 'Bugs in the brain'.)

- New kinds of electrical signals in the human brain have been discovered that separate us out from other species.

- The US Food and Drug Administration has approved Spravato ('ketamine' – the party drug and horse tranquillizer) as a nasal spray for the treatment of depression in adults.

- Challenging the idea that brain death is final, Yale researchers revived the disembodied brains of pigs four hours after the animals were euthanized. A frightening study, with huge implications for ideas about consciousness.

- And while we're on consciousness: scientist have produced 'mini-brains' which may have the capacity to think. Mini-brains are neuron clusters 5–6 millimetres in size, grown in the laboratory, which organize themselves into brain-like structures. They raise the nightmare view of a disembodied brain conscious of being trapped in some never-ending cycle of pain and suffering.

- The US corporation Biogen has resurrected the drug aducanumab for approval as a treatment for Alzheimer's. This development has had a huge impact on pharmaceutical companies, many of which were disengaging from the search for neurological cures. To put this development into perspective, as yet we have no cure for any of the four hundred neurological diseases classified by medicine.

Some of this, I admit, makes uncomfortable reading. So, in the spirit of the advice presented in this book, if you find some of these and other edgy findings disturbing, de-stress with some physical activity – research has shown that twenty minutes should do it. To give you a flavour of what is to come in the remaining chapters, here are just a few of the suggestions they will make: declare war on the chair to give you the edge (chapter 2); chew gum for added brain power (chapter 4); take Vitamin B6 before jumping out of a plane (chapter 5); and have more orgasms – they're good for the brain (chapter 7). I hope these points, and all the other advice that follows, will give you a greater understanding of your own brain and how to nurture it. Your life depends on it.

Body and mind

IN 2004 TWO DEMOGRAPHERS, Giovanni Pes and Michel Poulain, identified an obscure region in Sardinia as the part of the world with the most centenarians. As they narrowed their search, they drew blue concentric circles until they captured the exact location of peak longevity. Seizing the opportunity, Dan Buettner, a US explorer and author, worked with Pes and Poulain to identify four other locations blessed with high longevity (not entirely easy: one of the main findings was a strong association between claims of longevity and poor record-keeping!). These were Okinawa, Loma Linda (California), Costa Rica and the Greek island of Icaria. All five of these areas – first called the 'Blue Zones' in the November 2005 edition of *National Geographic* magazine – shared nine common features. One of them was moderate, regular and prolonged daily physical activity. As a way of life.

In all these populations, the way of life involved walking for ten thousand steps every day – at the very least. As shepherds, goat-herds, farmers. No one ever joined a gym, had a personal trainer, worked out or ran marathons. In their unwitting wisdom, these people foreshadowed the health advice now considered the vanguard of scientific progress, designed to overcome the ill-effects of our modern lifestyles, imprisoned as we are by hours every day in the car and/or at the office. The demise of physical labour in all fields of occupation places us in conflict with our natural physiology – developed not just for the relatively recent Neolithic farming conditions of ten thousand years ago, but over 1.5 million years of human evolution, through which survival was to a large degree dependent on the ability to gather and to hunt. Almost as if obeying a law of physics, our Western, sedentary lifestyle has been plagued by degenerative ailments – obesity, diabetes, heart disease and dementia. And, as if demonstrating the same law from the opposite direction, in the Blue Zones dementia and other forms of mental decline are rare – about 75 per cent less frequent than elsewhere in the Western world.

Evolution, modern life and physical activity

In the early 1830s Sir Charles Lyell, a lawyer and Professor of Geology at King's College, London, put forward what was then an astonishing (and provocative) idea. Not only was the Earth many millions of years old – but it had been shaped by climate change no different from that occurring now.

Geology tells us that the most prevalent condition of the Earth is glacial, that is, covered in ice. The last ice age ended a mere ten thousand years ago. It had lasted 2.6 million years. For three hundred thousand of those years, anatomically modern humans had occupied the planet, evolving as 'hunter-gatherers'. Then came an epic change in climate that tipped us out of the ice age into the Neolithic era: the warmer weather made agriculture feasible, and farming was born. It arose simultaneously in several areas of the globe, most famously in the Middle Eastern 'Fertile Crescent' (modern-day Iraq), and then spread to Europe. With it came profound changes in our behaviour and the type and levels of our physical activity, as we moved away from nomadic existence over vast areas towards settled farming, though the herding of animals still required long daily periods of walking.

The hunter-gatherer epoch had earlier laid down the evolutionary template for our modern bodies. In conditions of food scarcity, our physiology developed to support prolonged periods of hunting, involving long, slow treks interspersed by intense killing activity; then eating voraciously while food was available; then resting, to store and conserve energy for the next hunt – or for evading predators or for seeking a mate. Interestingly, as our ancestors moved from a scavenging to a hunting lifestyle there was a reduction in what is called 'sexual dimorphism'. In other words, male and female become less different from one another – so that, for example, female physical sexual signalling involves no conspicuous change in bodily appearance. The principal beneficiary of this change was reproductive capacity: unlike other primates, the female human is fertile, and so receptive, throughout the year – a phenomenon described by behavioural physiologists as the 'ERV' ('Ever Ready Vagina'). This means, in essence, that human group survival is more resistant to the pressures of population reduction because females can have children throughout the year. But that increased fertility meant that to get enough food to support the increased numbers of children, they were even more dependent on hunting *and therefore on physical activity.*

This survival-driven shift towards high episodic levels of physical activity as hunters was a critical event in our evolution. Hunter-gatherers used mobility as a survival mechanism, no longer relying on scavenging the food remains left by other predators but seeking out their own prey over vast distances. It is estimated that a hundred-strong hunter-gatherer group would require between 500 and 700 square miles of territory to support itself. Hunting placed a high premium on our physiological capability to maintain the health of our organ systems – our cardiovascular system, our endocrine system, our musculoskeletal system. And our brains. The evolution of our brain was driven by movement, prioritizing awareness, response, reaction, adaptation. And so, as modern science is revealing, the brain's health is dependent on the embedded need for physical activity.

Essentially, our physiology has not changed for 1.5 million years. Evolution has moulded the primordial clay of our human frames; recognizing this helps to explain the impacts of our current more sedentary lifestyle on our physical and mental health, and how we can use this evolutionary context to improve our well-being today.

The Discobolus

There could hardly be a greater contrast than that between our ancestral way of life as hunter-gatherers and our modern Western lifestyle. The progression to modern life, if we can call it that, has been a sad story of physical and arguably mental decline. For example, while human hunter-gatherers from around seven thousand years ago had bones comparable in strength to modern orangutans, farmers from the same area over six thousand years later had significantly lighter and weaker bones, more susceptible to breaking. Researchers now believe that it is not dietary changes but reductions in physical activity over thousands of years that are the root cause of degradation in human bone strength. It is a trend that is said now to be reaching dangerous levels, as people do less with their bodies today than ever before. It is estimated that more than 20 million people in the UK are physically inactive. If that were not enough, in a survey commissioned by the online training app Freeletics in 2019, some 42 per cent of respondents said they didn't have the time to exercise and 56 per cent said that even if they did, working life made them too tired. Even worse, a shocking 41 per cent said they were too old to get fit – at as young as 40.

Absence of physical exercise is a risk factor for many physical disorders, such as obesity, high blood pressure, heart disease and diabetes. In turn, the presence of these degenerative diseases is a risk factor for brain health. For example, obesity currently affects 13–20 per cent of children and young people in Westernized countries. Obesity is associated with a multiplicity of harmful health effects, including diminished brain function in childhood, and increased risk of Alzheimer's disease and dementia later in life. It gets worse. Obesity is well known to increase the risk of type 2 diabetes, which in adulthood is one of the strongest independent risk factors for neurological disease. In children and adolescents, recurrent high blood sugar (hyperglycemia) is associated with slower brain growth and reduced grey and white matter volumes, suggesting that carefully regulating metabolism early in life is important for long-term brain health.

The rise of the gym

On 20 July 1984, at the age of only 52, an icon of the American 'fame machine' and the author of the best-selling *The Complete Book of Running* died of a fulminant – explosive – heart attack while out jogging in his beloved Vermont. He was Jim Fixx, reformed smoker and exercise populist, the man credited with starting America's fitness revolution. Some might regard the ironic circumstances of his early demise as undermining his message about exercise – but they would be wrong. In his classic book published the following year, *Running Without Fear: how to reduce the risk of heart attack and sudden death during aerobic exercise*, Kenneth Cooper revealed that both Jim Fixx and his father were genetically predisposed to heart disease. By giving up smoking, losing weight and notably by taking up jogging, Jim Fixx had added some ten years to what would in any event have been a predictably short life. A 21-year study in Stanford University showed that elderly runners lead longer, healthier lives than their non-jogging peers. In the words of *Time* magazine, 'joggers live longer'.[1]

Since Jim Fixx's day, science has worked on quantifying the exercise we need, and government bodies on both sides of the Atlantic have published recommendations pitching our weekly ration of exercise at 150 minutes. Exercise helps to prevent long-term chronic illness, including cardiovascular disease, heart failure, obesity, high blood pressure and many other conditions. But it is not just the quantity of the exercise we do that's important. Science can now tell us not only how much exercise we should be doing but of what type, and how it should be varied by age. A critical factor is the level of exertion: experts differentiate between 'vigorous' – which includes running, cycling and energetic dancing – and 'moderate', which includes walking, pushing a lawnmower and ballroom dancing.

Public health advisory agencies such as the National Institute for Health and Care Excellence (NICE) in the UK spend a great deal of time critically searching the evidence on health-related matters in order to make recommendations. The NICE review process starts with a panel of experts, practitioners and representatives of the lay public agreeing on the key questions that have to be answered – in this case, for example, what age groups, what intensities of effort and what activities should be considered. Then a literature search is carried out. In the case of physical activity and exercise, NICE reviewed over ten thousand scientific publications. A summary of the evidence (called an 'evidence review') is then prepared, and to this are added cost implications. The review is then considered by a topic-specific committee and a draft guideline is prepared, which goes out to stakeholders, such as the public and health bodies, for rigorous and inclusive consultation before being signed off by the NICE Guidance Executive. By that point, we can be sure that the contents are firmly based on an exhaustive volume of evidence, on a scale well beyond the scope of this book to examine: a veritable goldmine of information. In the case of exercise, the recommendations, as adopted by the National Health Service, are summarized in box 2.1.

> **BOX 2.1: SUMMARY OF NHS RECOMMENDATIONS ON EXERCISE, BY AGE GROUP**
>
> **Under-5s**
> - Physical activity should be encouraged from birth, particularly through floor-based play and water-based activities in safe environments.

- The amount of sedentary sleep (sleeping while restrained or sitting) should be kept to a minimum.

Age 5–18
- At least sixty minutes of physical activity every day – this should range from moderate activity, such as cycling and playground activities, to vigorous activity, such as running and tennis.
- On three days a week, these activities should involve exercises to strengthen muscles and bones, such as swinging on playground equipment, hopping and skipping, and sports such as gymnastics or tennis.

Age 19–64
- At least 150 minutes of moderate aerobic activity, such as cycling or brisk walking, every week.
- In addition, on two or more days a week do strength exercises that work all the major muscles (legs, hips, back, abdomen, chest, shoulders and arms).

Over-64s
- Aim to be physically active every day. Any activity is better than none. The more you do the better, even if it's just light activity.
- On at least two days a week, do activities that improve strength, balance and flexibility.
- Do at least 150 minutes of moderate-intensity activity a week, or seventy-five minutes of vigorous intensity activity if you are already active, or a combination of both.
- Reduce time spent sitting or lying down, and break up long periods of not moving with some activity.

This advice is certainly helpful, but there is one glaring omission: at a time when we are all living longer and neurodegeneration is one of our greatest health fears, *there is little or no advice on the benefits of physical activity to brain health.*

What we know about physical activity, exercise and brain health

One of the first ever experiments on physical activity and the brain was carried out some forty years ago in Austin, Texas, by Waneen Spirduso, a visionary researcher at the University of Texas, now 83 years old and still practising. Professor Spirduso tested four groups of men (old and inactive; old and active; young and inactive; and young and active) for their reaction time and their movement time in specific activities. Reaction time is one of the best age-sensitive psychological and 'motor' performance measures. Her experiment gave rise to some unexpected findings. First, she noted that the performance of active older adults

An active lifestyle

(in simple reaction tests and movement tasks) was substantially better than that of inactive older adults. Second, and even more surprising, the performance of older active adults was similar to that of younger but inactive adults. In other words, a lifestyle of physical activity appeared to be *more important than age itself* in determining brain performance. Consider what this implies. The mature middle-aged executive who habitually exercises is likely to perform just as well as, or even out-perform, younger, unfit colleagues. This and other evidence suggests that ageing may be slowed down – and some would argue even reversed – by a lifestyle of regular physical activity.

This finding has been verified by many subsequent experiments. Spirduso's pioneering work was corroborated in 2007 by a 'meta-analysis' (a study reviewing and collating the numerical results of multiple previous studies) of the effects of physical fitness on the cognitive performance of older adults. By analysing eighteen longitudinal studies carried out between 1966 and 2001, two US scientists attempted to answer the question: 'Does aerobic fitness training enhance the cognitive vitality ['sharpness'] of healthy but habitually sedentary adults?' The answer was an unequivocal 'yes'. Their conclusion was that aerobic fitness training had a robust beneficial influence on the thinking skills of sedentary adults. Such fitness training was found to improve thinking skills of *all* types – decision-making, speed of thinking and memory. And there was great news for women – fitness-related improvements were larger in all female groups than in all male groups. In summary, the cardiovascular improvements gained from aerobic activity *appear to turn back the clock*. Science is just beginning to reveal how this works. What seems to happen is that aerobic exercise changes the human genome (our complete set of DNA) through a process called DNA methylation. Essentially, this means prompting chemical changes that slow down the ageing of our DNA or even reverse it. Further, evidence from cross-sectional, longitudinal and intervention studies with healthy older adults, frail patients, and people suffering from mild cognitive impairment and dementia all suggest that physical exercise is a potentially powerful non-pharmaceutical intervention to prevent age-related cognitive decline and neurodegenerative disease. Even imaging studies show positive responses to exercise training in the structure of the brain. For example, in 2011, a report for the US National Academy of Sciences cited a study of 120 older adults showing that aerobic exercise training increased the size of the hippocampus, leading to improvements in memory. One year of exercise training increased hippocampal volume by 2 per cent, *effectively reversing age-related loss in volume by one to two years*. From late middle age, most of us lose 1–2 per cent of our hippocampal volume every year, and researchers have found that this loss is associated with a reduction of memory capacity. In effect, persistent aerobic exercise protects our neural function – something we should all want to do.

How exercise exercises the brain

There is overwhelming evidence that aerobic exercise has beneficial effects on the brain, including improved mood and cognitive function (thinking skills). Now we are beginning to understand *how* these effects occur.

For many years it was scientific dogma that we are born with a set number of brain cells and that from the age of 25 we start to lose them – so after that it was downhill all the way! The good news is that scientists have now challenged this assumption, and there are studies indicating that we may be able to grow new brain cells throughout our entire life. This process is called neurogenesis, and is believed to occur in the hippocampus, an area we know is crucial to memory and the control of emotion. That's not all. The field of cognitive neuroscience has been transformed by the discovery that *exercise appears to promote neurogenesis in the adult brain*. And it gets even better. It is now clear that voluntary exercise increases levels of a growth factor in the brain, a chemical called BDNF (brain-derived neurotrophic factor). BDNF maintains existing brain cells, stimulates the growth of new cells and promotes the formation of connections between them (synapses). All these processes contribute to brain plasticity, increasing resistance to brain injury and improving learning and mental performance. There is reason to believe that enhanced neurogenesis may be associated with improved cognition (mental processing), whereas a decline in numbers of new neurons may be linked to ageing and depression; while these are still unproven ideas, the findings already to hand reinforce the message that it is never too late to support brain health by increasing our levels of activity and exercise.

Furthermore, we know that exercise affects multiple brain areas and systems, from synapses and neurotransmitters to the development of new blood vessels and brain metabolism, including glucose control. There is a growing body of evidence that, contrary to received wisdom, the brain does indeed play a role in the control of glucose in the body, independently of insulin. The question may then be asked: 'Does exercise affect glucose control *via the brain*?' – to which the answer seems to be 'yes'. We now know that during intense exercise, immense energy demands are made on the brain, prompting the brain cells to switch from using glucose to using alternatives (such as 'lactate' – a common dietary sugar that we regularly consume). This suggests that exercise training conditions not only the muscles (and the liver) but also the brain. What is more, these processes are two-way: during exercise, the brain affects the body's tissues and those tissues affect the brain, delivering multiple reciprocal benefits. As an example, we know that exercise promotes release of a special protein called irisin from muscle cells, and circulating irisin helps to produce BDNF in the hippocampus. As we exercise, we become more resistant to physical stress. Irisin released from exercising muscle is probably the critical 'muscle-to-brain' signal which sets stress-resistant processes in motion.

These mechanisms are millions of years old. From an evolutionary perspective, intermittent running and food deprivation (involuntary fasting) were the *most powerful drivers of change in the brain*. It should not surprise us, therefore, that our brains respond to physical activity by *expanding capacity* and to inactivity by *reducing capacity* – to conserve energy. The message is clear: for optimal health and efficiency, our brains are totally dependent on activity. We shall now look at what we can do to promote brain health by exercise strategies.

But before we do, a few words on where we're starting from. All of us alive today are likely to live much longer than our forebears. According to the latest estimates, there are 10 million people alive now who will live to the age of 100. Achieving longer lives is one of the greatest success stories of our age. But at the same time, our modern lifestyle has transformed later life into an age of degeneration and disease, with rising incidence of diabetes, high blood pressure, heart disease and osteoporosis. We should be ageing naturally, healthily. It is falsely assumed that disease is a natural consequence of getting older. It is not. There are many fit older people who have no chronic long-term illness. There is a lot of evidence to show that physical inactivity in later life is a huge driver of age-related disease. As we get older – 40, 50, 60 – we become overnourished, overmedicated and oversedentary.

If you doubt this, take a look at figure 2.1, which shows our energy expenditure over the course of a lifetime, both at work and through activity. At the age of 45, we fall off a cliff. By the time we are in our seventies, we have become grossly idle, indolent and inactive. In the words of Professor Todd Manini of the University of Florida, 'Aging induces a high propensity to be inactive.'[2]

What can we do to reverse this trend? Well, the first thing we must do, on the basis of the evidence on brain health, is differentiate between *physical activity*, on the one hand, and *purposeful exercise*, on the other. Physical activity is movement that is carried out by the musculoskeletal system that requires energy. In other words, any movement is actually physical activity, even getting out of bed. *Exercise* is planned, structured, repetitive and intentional movement intended to improve or maintain physical fitness. Exercise is therefore a sub-category of physical activity. Both increasing our daily activity levels and taking part in purposeful exercise can reverse the trend of increasing physical inactivity with age.

As we have already seen in this chapter, we have inherited bodies built

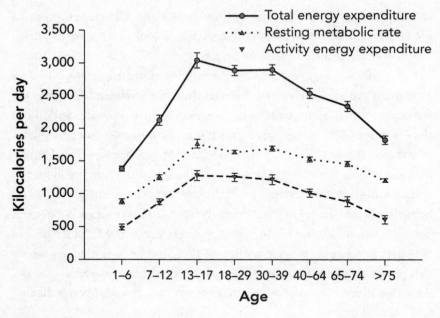

Figure 2.1: Energy expenditure at various ages

for habitual high levels of physical activity that are seamlessly integrated with vital processes in the brain. A physically active lifestyle is therefore consistent with substantial benefits for brain health. It's a current tenet of ageing research that, contrary to popular opinion, people can change their behaviour to become more physically active at any age and, in doing so, improve the prospects for their brain health. In short – and this is one of the core messages across every part of this book – it's never too late, at any age, to improve brain health. Epidemiological evidence shows that people who lead a physically active lifestyle have lower risk of cognitive decline, *from middle age onwards*. Other research shows that physical activity, as opposed to sedentary behaviour, is associated with less brain atrophy and, in turn, less damage to white matter (brain connections) as we get older. We know that preserving our white matter connections makes a sizeable contribution – about 15 per cent – to maintaining our intelligence levels across our life, from the age of 11 onwards. Improving and upgrading activity levels – 'getting moving' – starts with a change in attitude to incorporate more movement into all everyday activities. Changing attitude is not trivial. It is the essential precursor to changing behaviour, which is inherently intractable. We all resist changing how we do things. Some examples of behaviour change to get moving more might be: walk wherever possible instead of driving; take the stairs instead of the lift; if you drive, park farther away from where you are going; and think about engaging in hobbies and sports such as active yoga, dancing and gardening. For key practical tips on how to improve your level of physical activity, see box 2.2.

BOX 2.2: PHYSICAL ACTIVITY – PRACTICAL TIPS

- It's important to think about what will motivate you to exercise. Identify meaningful and enjoyable ways to increase and maintain physical activity, such as joining a social group, for example to go hill walking or walk your dogs.

- To stay motivated, consider doing physical activities with other people. Social aspects of physical activity can help inspire you to keep going and are independently beneficial to brain health.

- Make specific plans to move your body – think about when, where and with whom you will be physically active.

- Don't forget that whatever your age or current state of health, there are options to be physically active.

- Challenge yourself a little bit more over time. For example: (1) if you are not very active, start stretching and walking at a leisurely pace; (2) if you are already a walker or jogger, increase your pace or distance; (3) if you are an active runner, keep running and start strength/resistance training as well.

- Be patient and persistent – at any age, it takes at least a month for the body to respond to an increase in activity levels.

- Incorporate physical activity as a part of a comprehensive healthy lifestyle (more on this in other chapters) to help reduce the risk of cognitive decline.

Purposeful exercise (e.g. brisk walking or running, cycling, strength training, group exercise classes etc.) provides benefits for brain health, if it's moderate to vigorous in intensity. Examples of this would be walking, so long as it's fast enough to increase your heart rate; strength/resistance training (e.g. free weights, squats, lunges); and aerobic training which raises your heart rate (e.g. cycling, jogging, running, swimming laps, group exercise classes). Randomized controlled trials – the gold standard of scientific research – have shown that people who participate in purposeful exercise show beneficial changes in both brain structure and brain function. Though current public health recommendations focus on cardiovascular exercise, it's advisable to add to your 150 minutes of weekly moderate-intensity aerobic activity by doing muscle-strengthening activities on two or more days a week.

Practical tips for working out your own programme of physical exercise for beneficial effects on brain health are shown in box 2.3.

BOX 2.3: PURPOSEFUL EXERCISE – PRACTICAL TIPS

- Think about what exercises or physical activities you already do, and do more of them – unless you are already extremely active.

- Try new physical exercises and physical activities that you think you will enjoy.

- Do strength training and exercises that improve flexibility and balance in addition to aerobic exercise; a variety of physical activities is better than one kind alone.

- Think about getting advice from a physical trainer who can help you develop an exercise programme.

- Supplement your exercise by improving your daily physical activity levels, e.g. take the stairs instead of the lift; cycle or walk instead of using the car or public transport.

- Take advantage of time off from work, holidays and long weekends, such as bank holidays, to put extra effort into your exercise programme.

- But remember: if you have any health problems, always take the advice of a doctor before exercising.

But now – a real blockbuster. Even if we do all this, *there is one other, must-know secret that science has revealed and that you must act upon.*

Our fatal love affair with the chair

In 2005, Dr James A. Levine of the Mayo Clinic in Minnesota published a groundbreaking paper that added a key insight to what we know about the benefits of exercise. Its message? Being active is not enough – for the mind, the brain or the body. It is persistent sitting down which is the real killer. *Regardless of our level of physical activity.*

Sitting down is the silent addiction. Every day commuters sit for long periods in cars, trains and buses. Some 70 per cent of all workers spend seven hours, five or six days a week, in an office chair, getting up only for short breaks. At home, we sit to eat, to watch television or to read. Even when we go out, we sit in restaurants, bars, cinemas and the like – and almost invariably we drive, or sit on buses and trains, to get there and back. In the UK in 1950 there were 1.6 million cars. Now there are 38 million. It has been estimated that most people spend thirty of their 119 waking hours each week just watching television. Our nation – and Western culture more generally – has developed a love affair with the chair; and it's not doing us any good at all. How has this debilitating state of affairs crept up on us?

Between 1700 and 1750, Britain had an agricultural economy and the population remained pretty stable, at about 6 million. Both men and women engaged in long hours of physical labour, rising early, walking to their places of work and walking back home in the evening. Even 250 years later, at the turn of the twentieth century, life in rural England followed much the same pattern. The beauty of Flora Thompson's *Lark*

Rise to Candleford is that it depicts a lifestyle that most of us have never experienced. Thompson's appealing trilogy describes lives of hard physical labour, long-distance travel by foot, sparse and slow communication, and pleasure in small things.

In 1750 Britain imported 2.5 million pounds by weight of raw cotton, most of which was spun and woven by cottage industry in Lancashire. By 1787, raw cotton imports had risen to 22 million pounds, most of which was cleaned, carded and spun on machines. The industrial revolution had started. And with it the growth of the cities. Birmingham, population seventy thousand in 1800, had three hundred thousand inhabitants by 1850. The defining feature of the industrial revolution was migration into the cities, driven by opportunities to earn more and prosper in an age of great social mobility. Although for the majority life was still dictated by hard physical labour, the scene was set for continuing technological change and, with it, an increasingly sedentary working life. Since then our workplaces, schools, homes and public spaces have been increasingly engineered in ways that minimize human movement and physical activity. These changes have a deadly dual effect: *we move less* and *we sit more*. Our survival as a species was predicated on movement, and it was movement that drove the evolution of the body and the brain. This shift from a physically demanding life to one with few physical challenges has occurred only in the last 250 years, a tiny fraction of human existence. But during that time we have been slowly seduced to sit – and the effects have been dramatic.

The horrendous truth is this: even if we exercise daily, and beyond the approved recommendations, prolonged sedentary behaviour *wipes out the benefit of that physical activity*. Paradoxical though it sounds, we can simultaneously be categorized as *both* physically active *and* sedentary. Though first noted in the seventeenth century by the Italian physician Bernadino Ramazzini, it was not until Levine's research in 2005 that the negative health effects of sedentary behaviour became clear. He defined sedentary behaviour as long periods daily of sitting down, using up very low levels of calories. In scientific terms, any behaviour whose energy cost is less than 1.5 'METs' is sedentary. One MET ('metabolic equivalent of task') means your body is burning calories at a rate of 3.5 millilitres of oxygen per kilogram of body weight per minute. The actual amount of energy consumed will vary from person to person, depending on their weight. This MET value is a guide to the intensity of any activity. For example, sleeping is a 0.9 MET activity, sitting quietly is 1 MET, watching television is 1.3 METs, shopping 2.3, steady walking 3.6, leisurely cycling 4.0 and sexual intercourse 5.8. For runners, basic jogging is 7.7 METs while a flat-out four-minute mile will score you 23 METs.

Sedentary behaviour is tremendously destructive and plays havoc with our bodies. Data modelling predicts that the mortality rate increases by 2 per cent for every seated hour and by 8 per cent where the total seated time is greater than eight hours per day. Sedentary life, with its prolonged low levels of physical activity, increases the risk of almost every

Tennis champions, 1920

major chronic disorder, including heart disease, high blood pressure, diabetes, obesity and cancer. Put simply, there is another 'dose effect' here: the longer we spend in sedentary behaviour, the greater our risk of poor health. All the relevant studies show this. It is an inescapable fact of science. And it is true for people of any age, including children aged between 5 and 17. In Britain, less than 10 per cent of children spend more than one hour per day of their discretionary time in moderate to vigorous activity – most choosing to engage in screen-based 'activity'. There is a threshold for benefiting from physical activity, and it is quite high: daily moderate to intense exercise of 60–75 minutes is needed to compensate for long periods spent sitting. The inevitable result of a sedentary lifestyle is rising childhood obesity rates, decreased fitness – and, interestingly, lower levels of self-esteem, social activity and academic achievement. It gets no better as we get older. For adults, television viewing time, overall daily sitting time and time spent in cars are all associated with increased mortality from cardiovascular disease – and, in fact, all other causes. A recent study of fifteen hundred older adults found that those who were sedentary for ten hours a day and took less than forty minutes' moderate daily exercise had the bodies and health typical of people eight years older.

Though the science in this area is new, it can already tell us a lot about *how* sedentary behaviour damages our bodies. Recent studies have shown strong associations between adults' reported television viewing time – and, indeed, overall sedentary time – and very harmful health-related changes in the body. These changes include increased fat content (as measured by larger waist circumference), higher fats in the blood even when we are fasting (triglyceride levels), and higher blood glucose due to increased insulin resistance (the failure of the cells to recognize insulin). Increased insulin resistance is effectively a decreased tolerance for glucose and is a hallmark feature of ageing. These changes come about because, the longer we spend sitting, the fewer muscle contractions we make; and that means reduced breakdown of fats, reduced clearance of triglycerides, reduced clearance of glucose and less glucose-stimulated insulin secretion. All these changes are detrimental to health and are now known to be independent of the healthful effects of exercise. It's not just that exercise chemistry slows down when we are inactive; inactivity itself independently drives its own very harmful chemical changes in the body.

And, as if that was not bad enough, there are even more worrying findings: long periods spent sitting cause inflammation in the body to rise. As we saw in chapter 1, inflammation is a marker of ageing. The more inflammation there is in a body, the faster that body is ageing: so, the longer we spend sitting down, the more rapidly we age. Over a lifetime, the effects of all this sedentary behaviour are substantial. Those higher blood glucose levels cause inflammation that results in hardening of the arteries – including those in the brain as well as elsewhere in the body. There's more: it has been found that for every additional hour we sit in front of a screen per day, we increase by 7 per cent the likelihood of our having short telomeres (caps on chromosomes that protect DNA). Short telomeres are associated with shorter lives. Inflammation also affects the fat cells in the body so that stored fat becomes more resistant to weight loss – a process which has been called FATflammation. Cutting calories alone often isn't enough to deal with this. The answer lies in reducing inflammation induced by sedentary behaviour, improving activity levels over the long term and eating to reduce inflammation – a topic we'll consider in detail in chapter 5 on nutrition and brain health. Regular aerobic exercise such as that recommended above has been shown to reduce inflammatory biomarkers in the blood. For those who can manage it, long-term high-intensity training – such as running or rowing – raising the heart rate to about 80 per cent of its maximum for about an hour, two or three times a week, as well as everyday physical activity, is particularly beneficial in lowering inflammation.

Given the ravages wrought on our bodies by a seated lifestyle, it would be too much to expect that our brains would escape similar ill-effects – and so it proves. Studies have shown that sedentary behaviour is a predictor of Alzheimer's, and have calculated that about 13 per cent of Alzheimer's cases globally may be the result of inactivity. It is even estimated that a 25 per cent reduction in sedentary behaviour would reduce the incidence of Alzheimer's by about 1 million cases across the globe (there are approximately 50 million people suffering from Alzheimer's worldwide).

Other research has shown what the basis of this relationship may be – namely, that sitting down may well shrink the part of your brain tied to memory. Researchers at the University of California, Los Angeles, recruited thirty-five people aged between 45 and 75 and, using a self-reported questionnaire, estimated their average physical activity levels and the number of hours per day they spent sitting down over a week. Each person then had a high-resolution MRI (magnetic resonance imaging) scan which provided a detailed image of the medial temporal lobe or MTL, a brain region involved in the formation of new memories. The researchers found a strong association between sedentary behaviour and thinning of the MTL, which can be a precursor to cognitive decline and dementia in middle-aged and older adults. They also found that physical activity, *even at high levels*, was insufficient to offset the harmful effects on the MTL of sitting for extended periods. Though this study did not *prove* that too much sitting *causes* thinning of the MTL, it did show that those who spent more hours sitting down tended to have a thinner MTL – and that is a valuable indicator of risk.

Waging war on the chair

Thankfully, science also tells us how to counter the deadly effects of our sedentary lives. Think of it as a three-step process (see box 2.4 for a summary). The first step is get the message. Your mantra must be: 'Sitting still is bad for me.' Once you understand this message, it will motivate you to make simple, easy, straightforward changes that will have profound, longlasting effects. In short, declare war on the chair. The second step is to make a flexible anti-sitting plan – one you can stick to. Remember, what matters is not what you do on a single day; it's single things that you do every day. Start by monitoring your daily sitting time. Choose one day during the week and one at the weekend. Use a stopwatch (there's one on most mobile phones) to monitor how many hours and minutes you spend sitting on each of these days, and repeat over a few weeks. Once you have a pretty good average figure for how long you usually stay seated, aim to reduce it by 20 per cent in four weeks. And the third step is to *implement a daily plan of attack* to bring about this reduction. Start by making it a rule never to sit when you could stand. This sounds straightforward, but it's not easy to put into practice, for the simple reason that almost all of society is organized around sitting down. But many offices are converting to standing desks or desks of variable height. If standing is not an option, then you should do no more than one hour of continuous sitting, as an absolute maximum, whenever possible. Your rule should be to take a ten-minute movement break every forty-five or fifty minutes. It can be anything: going to the loo, getting yourself a cup of tea or coffee, walking up and down the stairs to raise your heart rate. Try to make breaks enjoyable and purposeful, and incorporate something you find interesting, such as chatting or making coffee or tea. The reason for this is that if you rely on willpower alone it will let you down – and these changes need to become permanent.

BOX 2.4: WAR ON THE CHAIR

Step 1: Understand the message
- Sitting is the silent addiction.
- Sitting is bad for your health.
- Too much sitting removes the benefit of exercise.
- Exercise does not counter the ill-effects of sitting.

Step 2: Make a flexible anti-sitting plan
- Monitor how much time you spend sitting down in a day.
- Aim to reduce it by 20 per cent in four weeks.

- Be flexible and forgiving: some days you will not meet your aim, but don't let this put you off.

Step 3: Implement a daily plan of attack
- Avoid sitting: stand wherever and whenever possible.
- Sit for no more than one hour at a time, if possible.
- Get up out of your chair for ten minutes in every hour.
- Make your movement time enjoyable and purposeful.
- If you can, work together as part of a group.

This 'enjoyment principle' is best illustrated by an experiment that was carried out at the world-famous sports science laboratories at Loughborough University. Two groups of young adults were tested on how long they could maintain a static seated position, unsupported, against a wall (anyone who has tried this will know it is total agony!). Then, just before lunch, they were split into two groups and asked to sit down and spend thirty minutes completing a questionnaire. While they were doing this, one group was faced with a plate of warm, delicious-smelling doughnuts which they were not allowed to eat. The other group – the control group – faced no such temptation. Straight afterwards, both groups were again asked to sit unsupported at the wall. Amazingly, more than half of the group that had been subjected to the inviting smell of doughnuts could not hold the position. They had drained their willpower, drawing on limited reserves of motivation. Psychologists call this 'the hindering effects of ego depletion'. The message from this? Willpower is important, but we don't have unlimited amounts of it. Using the reward of enjoyment in any activity will help to sustain it. Finally, don't go it alone: if you can, get together with others, be part of a group – family, friends, colleagues – all working together with the same aim. If you are a parent, you probably tell your children to get off the couch and play. But the kettle should not call the pot black. All of us at any age should do the same.

The Underwear Experiment

If you're still not convinced of the value of these simple changes, consider the famous 'Underwear Experiment' carried out in the Mayo Clinic in the USA. Scientists asked twenty people, ten thin and ten overweight, all of them self-describing as 'couch potatoes', to wear specially designed underwear. The top was either an undershirt or a Lycra bra, and the bottom was a risqué-looking pair of shorts with openings front and back – so the garment would not have to be lowered during the day. Why? So as not to disturb highly sensitive movement sensors embedded in the clothing. The participants wore their special sets of underwear night and day, for three periods of ten days, spread over a year. They could only take them off for fifteen minutes each day, to shower and get a fresh set. The participants' diets were also strictly controlled. It is worth saying that for their trouble they were paid $2,000 at the end of each ten-day phase, earning a total of $6,000 apiece. By the end of the study, which required a staff of 150, the researchers had collected 25 million pieces of data on each participant's movements.

And what did they find? Their conclusion was that all 'couch potatoes' are not the same. It was found that the overweight people had a tendency just to sit, while the lean ones had trouble keeping still and spent *two hours more a day on their feet, pacing around and fidgeting.* Amazingly, that difference in movement habits translates into about 350 calories a day, enough to produce a weight loss of 30–40 pounds in the course of a year – all without a single trip to the gym. They further concluded that it is the predisposition to be inactive that leads to obesity, not the other way around.

And what is the message of this research? I could not put it more eloquently than a recent article in *Psychology Today*:

> You probably can't get yourself to start fidgeting any more than a
> fidgeter can get themselves to stop. But if you're already a fidgeter,
> you (and others) can perhaps start to see it as a positive trait,

rather than one often associated with rudeness, impatience, and distractedness. And if you're not a fidgeter, you can learn to move more often and take frequent breaks throughout the day to walk around your desk, stretch your arms or otherwise move your limbs for a minute or two.[3]

At this point you, the reader, may justifiably ask: 'Are there any guarantees that if I follow all the advice in this chapter, I shall have a healthier brain?' There are two answers to this important question. The first one is that all the scientific research to date indicates that a consistent, longstanding and substantial change in lifestyle towards greater physical activity across many years will undoubtedly increase your chances of improved brain health. There are two important caveats: (1) increased levels of physical activity should be married to other vital lifestyle changes – what matters is not what you do on a single day but what you do every day, across a lifetime; and (2) we should be very circumspect in talking of 'guarantees' in science – owing to that trifling little inconvenience, which comes along with alarming frequency, called 'random events' that no one can predict. Having entered those caveats, it remains the case that our future well-being (and that includes brain health) is all about stacking up the numbers in your favour as high as you can – and physical activity is one of the biggest contributors to that pile.

The second answer relates to the Finnish FINGER study, discussed in chapter 1. This piece of research is a showcase for the potential we have to reduce the risk of cognitive decline by modifying our lifestyle. To summarize again, the FINGER trial is the first randomized controlled trial showing that it is possible to prevent cognitive decline among older at-risk individuals by intervention in several aspects of lifestyle. The researchers showed, after a two-year trial of over 2,500 people, that changing certain aspects of our lifestyle can reduce the risk of cognitive decline by 30 per cent over the course of a lifetime. The strength of the study is that it looked at a wide range of factors, including diet, vascular risk – and, notably, physical activity.

Again, though, we have to enter a caveat: while its findings are salutary and highly positive, the FINGER study is *only one piece of research*; the evidence on which to make recommendations for brain health is incomplete and still evolving. Having said that, despite not being able to tell us with absolute certainty which types of exercise are better than others, the body of research indicates that people who are less active can benefit their brain health by becoming more active. Furthermore, current evidence indicates that individuals who lead inactive or sedentary lifestyles stand to benefit more from starting an exercise programme than those who are already physically active do from taking up new forms of, or doing more, exercise. The reason for this lies in the scale and increased rate of change in new starters – especially the reductions in inflammation to the linings of the blood vessels. But even though there's insufficient research at the moment to tell us precisely what the optimal type of exercise for brain health is, there's certainly enough evidence to recommend that people should start thinking about the type, frequency, duration and intensity of exercise they currently get, either at work or in leisure time, and consider how they could increase or vary that. So the answer is YES: we *can* conclude that physical activity, including purposeful exercise, has a positive impact on brain health.

Back to the Blue Zones

The centenarians of the Blue Zones would no doubt smile at our preoccupation with exercise – with marathon running, exercise classes, personal trainers and gym membership. Most of them reached 100 years of age in traditional occupations such as goat-herding, farming or shepherding, walking many miles each day, with an elevated but not excruciatingly high heart rate – and little decline in their mental faculties. Their lifestyle drew them into close proximity with their families and their communities. Today, the enforced inactivity of our lifestyles reduces our confidence, saps us of energy, makes us irritable or bad-tempered and drains us of the motivation to spend time with our families and others – who are often distant as work uproots us. Poor levels of physical activity are only one part of an unhelpful modern lifestyle that marginalizes many of us and generates a lot of loneliness. We will deal with this issue in much more detail in chapter 6; but next, we will take a look at something you may not know you have: your 'second brain'.

Godzilla brain

IN ONE CORNER, JUST two men. One, a *Jeopardy!* winner seventy-four times in a row – Ken Jennings; the other, Brad Rutter, another veteran of the US television quiz show with $3.25 million of lifetime winnings to his name. In the opposing corner, not a man but a machine: Watson, a Linux-based natural language processor, the result of four years of development by twenty-five of IBM's best engineers. On 14 February 2011, these disparate opponents locked horns in the strangest ever episode of *Jeopardy!* – a kind of reverse general knowledge quiz, in which contestants are given clues in the form of answers and have to think up appropriate questions. Watson won, *PC World* magazine declaring the victory a 'shellacking'.

Behind Watson were ninety IBM Power 750 Express servers powered by eight-core processors – four in each machine, so thirty-two processors in all, yielding a total processing capacity of 80 teraflops. A teraflop is one trillion operations per second. Watson also had 15 terabytes (TB) of memory. Its inventor Tony Pearson, citing the futurist Ray Kurzweil to the effect that the human brain can hold about 1.25TB of data and performs at roughly 100 teraflops, declared: 'Watson is 80 per cent human.'

In fact, Watson was very far removed from being human. All it had to deal with was the single question posed to it at any one time. The brains of its only too human competitors were handling multiple other death-defying threats, as well as fighting their nerves – balancing salt and water levels, breathing, regulating heart rate and blood pressure, maintaining balance, controlling the subconscious. A few years later, Watson morphed into Debater, the first artificial intelligence (AI) system that can debate complex topics. With a database of 10 billion sentences, Debater was put to the test in 2019, debating against World Debating Championships Grand Finalist and 2012 European Debate Champion Harish Natarajan. Debater lost. So even this latest IBM offering cannot rival the capacity of humans in simultaneously performing a wide variety of tasks that require creativity, common sense, language and emotional empathy. We are still way better than machines at tasks that cannot be digitized or described by algorithms. Today's computers still need to be told exactly what to do and are only just beginning to learn by themselves, notwithstanding Stephen Hawking's assertion that AI is the biggest threat to the survival of humanity.

But here's the real clincher – energy efficiency. Current computing operations are underpinned by vast resources. IBM's Watson consumed about 750,000 watts of power. The UK's largest supercomputer, at the Met Office in Exeter, consumes as much energy in twenty-four hours as the city itself. The human brain runs on only 12 watts – one-fifth of the energy emitted by a 60-watt light bulb. That said, the brain drains the body of energy. It is gluttonous – to a far greater degree than that of any other primate. In our evolutionary past, that posed a considerable problem for humans. Of all the concerns of the natural world, acquiring energy is the most challenging. It drives the survival of the fittest. So how did humans solve this most critical of dilemmas?

Feeding the brain

Possibly as far back as a million years ago, a new technology revolutionized human culture and human evolution. This technology dramatically increased the availability of dietary energy to our human forebears. And in so doing, it precipitated two crucially important evolutionary developments: a massive increase in the size of the brain and an equivalent reduction in the size of the gut. We are talking about *cooking*.

In particular, *cooking meat.* This was a critical factor in the development of the human brain. First, it improved the availability of metabolic energy. The brain is a very expensive organ in terms of energy, representing only 2 per cent of the weight of the human body but accounting for 20 per cent of its total energy. By contrast, the brains of other primates, like chimpanzees and apes, consume about 8 per cent of the total. Meat is very dense and offers a lot of energy, but is difficult to digest in its raw state. Cooking breaks down the tissues and makes digestion easier, rendering the energy more rapidly available. Second, it enabled the human gut to shrink. No longer burdened with the need to handle high volumes of low-energy plant foods, humans evolved to have a relatively small total gut for their body size compared to their plant-eating ancestors. If we examine the remaining hunter-gatherer communities of today, more than 70 per cent of them rely on the consumption of meat for over half their dietary energy. Indisputably, meat has played a starring role in the evolution of our brains. But as much as we are omnivores, we are also 'coctivores' – adapted to eat cooked food – right down to the configuration of our teeth.

Cooking contributed to human evolution in two other ways: by facilitating the successful migration of humans to every habitat of the planet; and by promoting our vital social cognition, in which we are supreme among animals. The key to successful migration is the ability to exploit new food environments – and here the ability to cook gives you a huge advantage. If you can cook what you find, you can turn a lot of what would otherwise be indigestible into food. Whatever humans found, they could cook. And eat. In human social life, cooking became – and remains today – a central, culturally binding activity. Which conveniently brings us to the other half of the hunter-gatherer lifestyle: gathering. If hunting provided an indispensable source of energy, then gathering – of herbs, fruit, roots and seeds – provided another indispensable dietary asset, that of diversity. There is evidence that we were eating grains at least a hundred thousand years ago – easily long enough for our gut to have evolved since then to accommodate them. It's natural to eat whole grain. The Palaeolithic – stone age – diet was hugely more diverse than the narrow Western diet that is the standard fare of millions today. Its diversity provided vitamins and micronutrients, such as minerals, all of which fuelled the development of brain metabolism. But gathering wasn't just about providing extras: as anyone who does it will tell you, hunting is a precarious business. More often than not it is unsuccessful. The gatherers, typically women, provided the fallback calories during difficult times.

This evolutionary journey was not a one-way street, with everything working for the development of the brain. To meet the high energy cost of developing the brain, it was the gut alone – of all the tissues in the body – that could compensate by reducing in size as the quality of the diet improved in terms of calorific content, micronutrients and energy availability. This cannot be said of any of the body's other organs, whose size is non-negotiable, determined by what they have to do in the body. So, as much as the gut developed the brain, so the brain in turn developed the gut – and, what's more, a dedicated line of communication between them.

The second brain

The Japanese film monster Godzilla has long been portrayed as an unstoppable force – hence one of his titles, 'King of the Monsters' – but he has a hidden and fatal weakness. Godzilla has two brains, and one of them is in the most unusual place, where his tail meets his torso: and that was the key to his defeat.

We too have another brain within us, as well as the one in our skull – in our gut. It has 500 million neurons, twenty different types of nerve cell and multiple complex 'microcircuits', and is the size of a domestic cat. This is the enteric nervous system – our own Godzilla brain. It not only monitors the entry of food into the body, registering its taste, texture and condition; it also regulates digestion – the breakdown of food, its passage along the gut, the absorption of breakdown products

Godzilla, the beast with two brains

and the expulsion of what is left. As we all experience, some of this is less than satisfactory! And throughout these processes it is in constant and vital communication with the brain in our head. We are constantly making conscious decisions about what we feel is going on inside us and how we behave socially. Even the four taste sensations identified by the human tongue have been honed to serve a survival function: salt – indispensable to the electrical function of the nervous system; sweet – a 'barometer' of calories; 'bitter' and 'sour' – warnings to protect us from plant toxins. Thus taste, reward and energy are all carefully wrapped in a neat survival package. Our behaviour around food and eating is a complex area of psychology and physiology that lies beyond the scope of this book. It's enough for present purposes to say that the gut and brain together have a very sophisticated control system for the consumption and regulation of food intake. It stretches from the furthest reaches of the gut (the colon, where the digestion process concludes) to the brain stem (satiety or satisfaction centre), and from the subconscious limbic system (indulgence) to the cerebral hemispheres (conscious control of eating behaviour). There is also research evidence that the brain has what some dieters may regard as a pernicious and somewhat perverse feature: a body weight thermostat which defends a set weight of the body.

'Gut wrenching.' 'I feel it in my gut.' 'Stop belly-aching.' 'Butterflies in the stomach.' A plethora of everyday terms pays testimony to the commonly felt system of communication between the Godzilla brain in our gut and the human brain in our head. Scientists call it the 'gut–brain' axis, and it is essential to our health and well-being. The main nervous pathway from brain to gut is the vagus nerve. This was named from the Latin word *vagus*, meaning 'wandering' or 'straying', because it serves so many internal organs. But some 80–90 per cent of the individual nerve fibres in this huge cranial nerve – the largest in our body aside from the spinal cord – are *dedicated to messages between the brain and the gut*. And there's more surprising news: only 10 per cent of these interconnecting nerves are dedicated to delivering messages *from the brain to the gut*, whereas 90 per cent are dedicated to messages *from the gut to the brain*. The Godzilla brain is telling us what is going on! Never mind the eyes, the ears, the fingertips or skin – it turns out that the gut is the largest sensory organ in the body. The activity going on there is so important that there is a second-by-second messaging system relaying information from the gut to the thalamus. Once the messages reach a critical mass, they are conveyed to the cortex, generating feelings which make us feel well or unwell – and fifty shades of grey between, contributing significantly to the state of our mood.

And, in a curious act of biological irony, one of the brain's most important messenger molecules, the 'happy hormone' serotonin, is *not* made exclusively in the brain. Amazingly, some 90 per cent of all serotonin originates in the gut – from the trillions of bacteria known as the gut microflora or microbiota. Serotonin is a major factor in our mood states. Bad food, bad mood.

All this complexity compels us to ask: 'What is so important that it requires such a potent system of interaction – to the point where we evolved a Godzilla brain?'

Getting our energy

The acquisition and mobilization of energy is the single biggest driver of all life in the natural world, including ourselves. Certain organisms called autotrophs ('self-feeders') are the only organisms that can trap 'free' energy' and use it to build living material. They are fundamental to the food chains of all ecosystems in the biosphere, of which we are a part. These autotrophs take energy from the environment in the form of sunlight (green plants) or inorganic chemicals such as sulphur (bacteria). They use it to create energy-rich or 'highly reduced' organic molecules such as glucose – 'organic' meaning based on carbon atoms. Humans, by contrast, are 'heterotrophs': that is, we cannot directly harness energy for ourselves, so we have to acquire it either directly, by devouring autotrophs (such as green plants), or indirectly, by consuming other organisms (herbivores) that eat them. Largely, therefore, the food we eat consists of 'highly reduced' or energy-rich large organic molecules – carbohydrates, fats and proteins.

The primary purpose of the digestive system – the gut – has an elegant simplicity about it. We take in these large molecules and the gut breaks them down into molecules small enough to be absorbed into the blood, which then distributes them around the body so that our various cells can acquire the energy in them. The breaking-down process (digestion) is in chemical terms called 'hydrolysis' (literally, 'breakdown by water'). Even at our warm body temperature, this process would be too slow to meet our energy needs, so powerful catalysts called 'enzymes' are used to speed up the breakdown of the large food molecules. Enzymes are produced within the cells of the digestive system and released into the food in stages as it passes through the gut. The small molecules that result from the digestion process are transported to the cells of the body which 'burn' them using oxygen to release energy. And the organ that consumes more of this energy than any other is the brain.

And yet, for over three hundred years after anatomist Andreas Vesalius published the first accurate description of the gut in 1543, scientists had very little clue as to what went on there, let alone how it worked. As so often in science, it took a chance event to enlighten us.

A grim accident with a profitable legacy

From 1777 to the present day, Massachusetts has made most of its money from firearms. Revolutionary enterprise and ingenuity produced one of the most innovative and deadly weapons of its time – the 1816 .69-inch calibre flintlock musket. On 6 June 1822, it showed its power in a gruesome accident at the fur-trading post on Mackinnock Island that revolutionized not only our knowledge of human digestion but the practice of experimental physiology. And changed for ever the life of one Alexis St Martin.

With the musket loaded and the weapon safely at half-cock, what could possibly go wrong? As the hunter obliquely rested the weapon, little did he know that a slow-burning ember was creeping along the vent from the firing pan towards the main charge in the breech. As his friend and fellow hunter Alexis walked away from the muzzle, the charge exploded, sending a hurtling cloud of duck shot into his back. A searing slug of lead ball, spent powder, burned wadding, muscle, skin, bone and lung, the size of a man's hand, entered the chest cavity of the hapless Alexis St Martin. And exited via his stomach.

As St Martin lay there in a pool of blood and gore, no one thought he would survive. But he had the good fortune to be attended by a US Army doctor, one William Beaumont, well experienced in military surgery; and, over many months, the injured man recovered. The healing process, however, left an unusual legacy. A round fistula or open wound remained unclosed, connecting the stomach to the outside world. After eighteen months it had developed a sphincter, by which Beaumont could inspect at will the contents of the stomach. And so began a ten-year series of experiments in which the good doctor would lower, on a thread, food of all types into the stomach, later retrieving them and recording their condition. Meat, fish, offal, eggs, bread, vegetables, fruit – all were examined. And tasted, before and after. He also undertook the process in reverse, extracting fluid from the stomach at different times of the day and night, before, during and after eating and drinking. And so, by virtue of a ghastly accident and a diligent doctor, the scientific investigation of human digestion began.

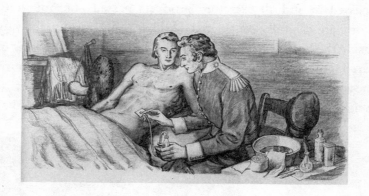

Surgeon William Beaumont and his patient Alexis St Martin

Beaumont's role in this unfortunate episode amounts to a miracle of medical science. It went beyond saving a life. It also began a vital journey towards understanding our own gut, a process which continues to evolve today. Described by the best-selling author Giulia Enders as 'the body's most under-rated organ',[1] the gut is a wonderfully honed construction: the alimentary canal, 40 metres long from mouth to anus, has a huge internal surface area of 400 square metres and is served by complex accessory organs to aid digestion along its route – the salivary glands, the pancreas and the gall bladder. Our modern knowledge of the gut, food and digestion has been won by the painstaking efforts of real pioneers who risked their reputations and livelihoods to piece together the knowledge we now take for granted. And no one more so than Wilbur Olin Atwater.

It's all in the numbers: counting calories

Imagine being incarcerated in a metal and wood box, 7 feet long by 4 feet wide by 6 feet high. In this chamber you eat, drink, work, rest and sleep. It is ventilated for fresh air and you are kept pleasantly warm. You have a small folding bed, a chair and a table. All food and drink are passed in and all excretory products passed out. Your creature comforts are carefully considered – which is just as well, for you will be inside the box for days. The year is 1896 and you are but one of five hundred individuals who will take part in this experiment. Its purpose? To measure the energy content of food. It is to this fiendish device and the scientist who operated it – Wilbur Olin Atwater – that we owe our

Wilbur Olin Atwater and his respiration calorimeter chamber

obsession with counting calories.

The idea of a 'calorie' entered popular Western vocabulary with an article by Atwater published in 1887 in *Century* magazine. It was defined specifically as the amount of energy which would raise the temperature of 1 kilogram of water by 1 degree Celsius. Now, before we go any further, we need to clear up some confusion. In Europe, a 'small' calorie was defined in terms of heating 1 gram of water. You may think that this is an academic nicety, but it leads to hideous confusion between scientists and the public. It works like this: 1 Calorie (large 'C') is actually equal to 1,000 calories (small 'c'). Therefore 1 Calorie (large 'C') is sometimes referred to as 1 Kilocalorie or 1 Kcal. Either way, the energy values given on our food packaging are mostly in Calories (large 'C'), and that is why they are sometimes marked up as 'Kcals'. It is this number that is meant when we talk in everyday usage about 'calories'.

Thanks to the work of Wilbur Olin Atwater and other pioneering scientists, we now know the energy content of virtually every food on the planet, broken down into quantified servings, raw or cooked. And similarly, we know the energy requirement for every human activity (and I mean every activity!). These values are closely related to our health, our weight gain or loss, our diet and our activities – and, as we shall see, the health of our brain. To find out what your daily energy need is, by age and gender, all you have to do is to look up the appropriate database online. To save you time, the key numbers as currently prescribed by HM Government in the UK are summarized in box 3.1.

BOX 3.1: UK GOVERNMENT RECOMMENDED DAILY ENERGY INTAKE

Age	19–64		65–74		75+	
Sex	Male	Female	Male	Female	Male	Female
Kcals	2,500	2,000	2,342	1,912	2,294	1,840

And, to reiterate a point made earlier in this chapter, fully 20 per cent of this daily energy requirement is delivered to the brain – if you are a 40-year-old woman, about 400 calories; for a man of the same age, 500 calories. In each case, that's the size of a typical modern-day breakfast. Your brain has one meal a day to itself!

It's commonly believed that the Minnesota Starvation Experiment in 1944 was the first investigation into restricting daily calorie intake. It was not. In 1934 a milestone paper was published that remained largely ignored for fifty years. Its simple yet profound message was this: *laboratory rats fed a calorie-restricted diet live almost twice as long as their well-fed counterparts.* It was written by Clive McCay, a professor of biology at Cornell University who was to become known to the general public not for this finding but as an expert in wartime rations – as a result of which the 'Cornell Loaf' was named after him. Little did he know that the principle of calorie restriction would survive over eighty years of research and testing to the point where it stands out today as the dietary intervention most effective in extending lifespan and delaying the onset of age-related disease. This holds true in all organisms from yeast to fruit flies to worms to rodents to primates. And humans?

First, an important clarification. 'Calorie restriction' is not just another name for a weight-loss diet. Calorie restriction means reducing average daily calorie intake substantially below what is typical or habitual but without incurring malnutrition or deprivation of essential nutrients. Short-term human trials have shown clearly that calorie restriction can improve many of our vital health signs, such as our body weight, blood pressure, and levels of blood sugar, insulin, blood cholesterol and triglycerides. It also reduces inflammatory markers such as CRP in the blood *and in the brain.* In these studies, participants were asked to reduce their calorie intake by between 20 and 30 per cent. Unwittingly, they generated one of the most important findings: namely, that the vast majority of humans cannot sustain the level of calorie restriction required to elicit all the potential favourable outcomes. Huge willpower is required, and the feelings of hunger and deprivation are, apparently, excruciating, miserable and completely demotivating. The people in these studies managed on average a reduction in calorie intake of only about 10 per cent.

So the search began for a *calorie restriction mimetic.* In other words, could we work out what happens in the cells when we restrict calorie intake and then identify a therapeutic molecule which would replicate this effect, extending our lifespan and reducing the risk of illness, *without our having to eat less*? Sounds too good to be true? Amazingly, scientists have now found several candidate molecules, such as aspirin, curcumin, rapamycin, metformin and resveratrol – and clinical trials have been started. There are currently over two thousand of these trials running, and it may interest the reader to know that *we can already buy products derived from these substances online.*

So: does reducing calorie intake improve the health of our brain? As we have already seen, in energy terms the brain is expensive, consuming 20 per cent of the body's energy. This equates to about 400 calories if you consume 2,000 calories per day – more if you work the brain harder, though thinking harder is not a good way to lose weight because the increases are trivial, in the order of only 20 calories per day, compared to the brain's baseline gluttonous consumption. To begin by taking the question from the other end, we certainly know that overeating will increase the risk of impaired brain function. In 2012, a US study of more than twelve hundred adults aged 70–90 showed that high calorific intake (more than 2,143 Kcals per day) from mid-life onwards *doubles* the risk of memory loss in later life compared to a reference group consuming less than 1,500 Kcals per day). We are pretty sure, moreover, of *how* consuming excess calories damages brain health – mainly by overloading the cells of the brain with what are called 'free radicals' or oxidants. These are electrically charged particles which have excess electrons (e.g. surplus oxygen) and cause 'oxidative stress' by seeking out and attacking substances in our cells, even our DNA. Every day, we experience over thirty thousand attacks on our DNA by free radicals, including in the brain. Free radicals are among the normal products of our cells; but too many of them will wreak havoc with our metabolism – and when we overeat, the mitochondria (energy producers) in our cells release a lot of damaging free radicals. So eating excessively is definitely not a good idea. Conversely, there are foods which contain many 'anti-oxidants' that combat free radicals, and a healthy diet will provide these. We'll come back to this point in chapter 5.

To return to the question posed above, there is indeed abundant proof that reducing calorie consumption, even moderately – say by 10 or 11 per cent – confers multiple benefits on the brain. Restricting the numbers of calories we eat is anti-inflammatory, reduces oxidative stress, facilitates 'synaptic plasticity' in the brain and promotes neurotrophic factors (such as BDNF, discussed in chapter 2 in the context of exercise), which stimulate brain cell growth. In summary, it prevents age-related damage to our brain cells. The message is relatively simple: eat less – and if you want to know how much less, try to feel hungry at least once a day!

By way of supporting evidence, let's take a look at the Munster Study. This was carried out in the German city of Munster in 2008, when a group of fifty healthy older adults with an average age of 60 were divided into three groups. Group 1 were asked to cut their calories by 30 per cent (most managed about 10 per cent); Group 2 increased the proportion of unsaturated fat over saturated fat in their diets; and Group 3 was the control group, making no dietary changes. At the beginning and end of the three-month experimental period, all the participants underwent memory tests. At the end of the three months, the calorie-restricted group – and only this group – increased their scores by about 20 per cent. This group were also found to have lower insulin levels – a characteristic feature of calorie restriction – and lower levels of inflammation (CRP) than the other groups.

What are we to make of these results? The first conclusion would be that high calorie intake is not only bad for your heart, it's also bad for your brain; the second, related, conclusion, that reducing calories benefits brain health. Though the Munster Study was small in size, its findings are supported by many other research projects worldwide, including studies of the 'Okinawa diet', which involves leaving the table early when only about 80 per cent full. Followers of the Okinawa diet thus eat about 20 per cent fewer calories than the Japanese average. As we saw at the beginning of chapter 2, Okinawa, with its traditional way of life, is one of the world's privileged Blue Zones, with high longevity and low prevalence of Alzheimer's disease.

A summary of what we know about calorie intake and the brain is shown in box 3.2.

BOX 3.2: CALORIE RESTRICTION AND BRAIN HEALTH

- Calorie restriction of 20–30 per cent is the dietary intervention known to be most effective in extending lifespan and delaying the onset of age-related disease in almost all living things.

- Calorie restriction at this level cannot, however, be recommended as a dietary regime; it's just not sustainable.

- In practice, most people can sustain daily calorie restriction of about 10 per cent of their habitual daily intake. We can call this 'calorie reduction'.

- Calorie reduction of about 10 per cent confers multiple general health benefits in humans – improving body weight, lowering blood pressure and levels of blood sugar, insulin, blood cholesterol and triglycerides, and reducing inflammatory markers in the blood.

- Reducing the amount of calories we eat has been shown to improve cognitive function, e.g. memory and learning.

- By contrast, overeating (more than 2,000 calories per day) increases the risk of impaired brain function.

Eating like our ancestors

Our typical eating pattern of three meals a day plus snacks flies in the face of our evolutionary history. Over 1.5 million years of eating sporadically, our brains evolved in response to what we might call 'intermittent fasting'. The evolutionary pressure of episodic food shortages developed a brain whose activities were sharpened when we were hungry or even starved. At rest, we deplete the energy supplies of our liver in about twelve hours of non-eating. Most of us are ravenous by then. Extend that to twenty-four or even forty-eight hours and our brain has no glucose left. We are running on ketones – the end result of 'ketogenesis', the chemical breakdown of fat stores in our body. Add in a period of a vigorous activity like hunting and we reach that point in less than twelve hours. Technically, a ketone is a simple organic molecule which has at its core a single carbon atom connected to a single oxygen atom by a powerful 'double' bond. Around that core, other chemical groups are connected which make the ketone quite difficult to break down to release its energy. In other words, ketones resist oxidation. Powerful enzymes in the cells of the brain are needed to oxidize ketones for their energy. Ketone is the evolutionary alternative energy supply to glucose – but in modern life we deny it to the brain. Over the course of our evolution, our brain has thrived not on comfort but on shock treatment, on challenges – and its response was to cope under severe pressure of enforced food shortages. So an obvious question arises: would the performance of our brains – our cognition – be enhanced by changing from three meals a day plus snacks to an eating pattern based on intermittent fasting?

This theory has in fact been tested – in animal studies. Intermittent fasting in small mammals has been shown to confer a number of benefits. It protects the brain against stress, increases the production of neurotrophic factors such as BDNF and promotes the growth of brain cells in the hippocampus. In response to the challenge of fasting, the brain also cleans up its act, conserving its resources and recycling dying and damaged cells. When feeding begins again, the brain returns to its growth phase – making lots of proteins and forming new synapses, all part of what we call neuroplasticity.

Controlled trials aimed at determining whether intermittent fasting improves cognition in human subjects have not yet been conducted. However, many readers will recognize the underlying chemistry of intermittent fasting. It is in essence a 'keto-diet' (high in fat, low in carbohydrate). When anyone eats a keto-diet, they burn fats instead of carbohydrates, and this results in free ketones being excreted in their urine: that is, they are in what we call a state of 'ketosis'. Has anyone looked specifically at a keto-type diet and brain performance? The answer is yes. The results of a study published in the *Journal of Alzheimer's Disease* in 2019 showed that it might be beneficial for people who are showing early signs of cognitive decline. This was a small pilot study conducted at Johns Hopkins University, Baltimore, where fourteen older adults with mild cognitive problems were put on a modified Atkins (keto) diet ('MAD'), while five other people went on a standard 'healthy' diet. The 'MAD' dieters were asked to eat less than 20 grams of carbohydrates a day, with no restriction on calories (we normally eat about 250 grams of carbohydrates a day). Both groups had their urine checked for ketones, to determine whether they were sticking to their diets. The presence of ketones in urine is proof of energy being derived from burning fats as opposed to carbohydrates. Now – the proof of the pudding. Did the ketone group do any better on tests of mental performance? The keto-diet fans among you will be delighted to know that the 'MAD' group scored significantly better than the other group on memory tests – and, what's more, the scores peaked with the peaks in ketones.

Think back to our ancestors. Between hunting trips, ketosis was a natural condition, and the brain had evolved to thrive when glucose

Fasting

was unavailable by oxidizing ketones for energy. There is, therefore, a case for intermittent fasting as a means of optimizing the brain's functioning by forcing it to burn ketones, its evolutionary natural standby fuel. However, it is worth entering a note of caution here: it has to be said that direct evidence from large-scale clinical trials is still awaited as such testing is still in its early stages. We do already have, however, a review of the evidence to date, published in the *New England Journal of Medicine* in 2019 by two US professors, Rafael de Cabo of the National Institute of Aging and Mark Mattson of Johns Hopkins University. They found intermittent fasting to have positive effects in cases of cancer, cardiovascular disease and diabetes, and in several neurological conditions – epilepsy, multiple sclerosis, Parkinson's and Alzheimer's – though many of the studies they reviewed involved non-human animals only.

Calorie reduction and brain disorders

So far in this chapter, we have looked at diet and nutrition in relation to the healthy brain – and how to keep it that way. What does the evidence say about brain disorders, like Alzheimer's disease? The reader should remember that the degenerative process underlying Alzheimer's can begin, imperceptibly, *as early as age 35*.

Multiple animal studies have found that sustained intermittent fasting slows cognitive decline. Small mammals undergoing this type of diet had better cognitive function, lived longer and, most importantly, had less plaque buildup (a clear signal of neurodegenerative changes) in their brains. Though the precise underlying mechanisms are unknown, there are clues – intermittent fasting is thought to increase resistance to attack by free radicals (oxidative stress) and switches the body from glucose to ketone metabolism (as in the ketone diet).

We don't know, at the moment, whether intermittent fasting prevents neurodegenerative diseases of the human brain. It may well do – we just don't have the evidence yet. There is currently a very promising study being carried out in Italy by Dr Valter Longo, Director of the Longevity Institute at the University of Southern California and an expert in fasting and the ageing process (also founder of the ProLon® diet, which very cleverly mimics fasting). The Italian study is looking at the effect of the ProLon diet on people with either Alzheimer's or mild cognitive impairment. We will have to wait to see if there is an unequivocal case for adopting intermittent fasting; but in principle, the evidence available to date supports the idea that it is beneficial for the brain.

Eating, not eating and looking after your brain

All the evidence shows that overeating is bad for the brain. On the plus side, actively reducing our calorie intake, and intermittent fasting to switch the brain periodically to ketones, appears to be *very good* for brain health. Coasting along in our comfort zone of constant eating and snacking is not such a good idea. It is better to 'shock' the brain by 'metabolic switching', that is, switching from glucose as the main energy source to ketones.

What, then, is the best way to put these ideas into practice? Here is a list of practical tips on how to adapt your eating habits to improve your brain health in the long term.

1. Don't try extreme calorie restriction, that is, reducing your intake by 20–30 per cent. There's not enough evidence to recommend calorie restriction on this scale as a dietary regime. Scientists still have much to learn about how 'true' calorie restriction and prolonged, acute fasting affect the general health and the brain health of people who are not overweight, especially older adults. We don't know whether eating patterns based on a 20–30 per cent reduction in calories are safe or achievable in the long run, but findings to date suggest they are not. And, as noted above, few people can actually make such a drastic reduction in calorie intake.

2. Having said that, *do* try to eat fewer calories. Evidence has shown that a reduction of 10 per cent is good for the brain – reducing inflammation and oxidative stress, and increasing neurotrophic factors (such as BDNF) which stimulate brain cell growth. A comparatively modest 10 per cent reduction in calorie intake is perfectly achievable; and it needs to be sustained, to become a way of life. One way of doing it is to follow the Okinawan principle of leaving the table slightly less full than we would like to be (in Japanese: *hara hatchi bu*). In modern society, it's very common for people to do the very *opposite* – that is, stay at the table until we feel *too* full. How many times have you and I done this? Once in a while, it will do no harm. The problem arises when it becomes a nasty habit.

3. Decide whether to have a 'normal' schedule of two or three meals per day or to fast intermittently (point 5 below), remembering the health caveats in this chapter (points 6 and 8 below). For example, if you are of healthy weight, or if you tend to lose weight easily, or if you are over 65 (provided your weight is within the healthy range), you could eat three modestly sized meals a day within the recommended calorific levels and one low-sugar snack (containing no more than 5 grams of sugar) with fewer than 100 calories – for example, a few unsweetened, non-salted, raw nuts. We will look more closely at diet in chapter 5.

4. There is good evidence that intermittent fasting protects your nerve cells and keeps them functioning better for longer. If you go for intermittent fasting, you will probably lose weight, but remember that weight loss is not the main driver of the health benefits observed in pre-clinical and clinical studies. Rather, the key mechanism is 'metabolic switching': that is, getting the

brain to switch from glucose to ketones for its energy supply. This is a natural switch which is damagingly absent from modern life.

5. There are two ways of intermittent fasting:

(a) Time-restricted eating. This means narrowing the 'window' during the day within which you do all your eating to between six and ten hours, which will leave at least fourteen hours of not eating between dinner in the evening and breakfast the next morning. This enables your body to switch to ketones overnight. To make the process easier and more sustainable, ease yourself in over four months until you get to a six-hour feeding regime. One caveat here: if you are facing big mental or physical challenges and want to maintain performance, you should not be relying on ketones.

(b) 5:2 intermittent fasting. This involves eating normally for five days a week and then on the other two (non-consecutive) days greatly reducing your calorie intake to, say, 1,000 or even 500 calories, drinking tea or water throughout the day. This type of intermittent fasting is recommended by Professor Mark Mattson of Johns Hopkins University, who is also Chief of Neuroscience at the US National Institutes of Health. It requires real application to keep going, but you'll be encouraged by improved energy levels and feelings of greater well-being. There are many intermittent fasting apps on the internet to help you.

Box 3.3 summarizes how you can apply each of these methods progressively over a four-month period.

**BOX 3.3: THE TWO KINDS OF INTERMITTENT FASTING
AND HOW TO APPLY THEM**

Month	Time-restricted eating	5:2 intermittent fasting
1	10 hr window, 5 days/week	1,000 calories, 1 day/week
2	8 hr window, 5 days/week	1,000 calories, 2 days/week
3	6 hr window, 5 days/week	750 calories, 2 days/week
4	6 hr window, 5 days/week	500 calories, 2 days/week

6. Intermittent fasting is *not* recommended for young children, or for those who are underweight, have type 1 diabetes, have an eating disorder, or are elderly and frail.

7. If you don't actively reduce your calories, then do avoid eating in excess. Moderation, one meal at a time, matters a lot. Eating excessively is not only bad for the heart but bad for the brain. And remember the Okinawan rule, *hara hatchi bu*: try to leave the table slightly hungry. A real challenge!

8. Before making any significant changes to your calorie intake, talk to your doctor.

We are a very clever species. We solved the problem of energy acquisition by a skilful double strategy. By cooperative hunting and gathering, we captured prey which yielded high-energy foods – meat and fish – and garnered a varied menu of plant foods, which provided standby calories and a wide range of micronutrients. By cooking, we rendered the energy content of food easily assimilable in whatever environment we exploited. Our precarious existence as hunters delivered a precious mechanism in the brain enabling us to survive in times of scarcity, forcing it to switch from glucose to ketones to supply the energy we needed when our prey eluded us. Thus we evolved our natural pattern of eating behaviour: eat when you can, hunker down when you can't. But now, in our food-rich modern economies, the ancient compulsion to eat food when it is in front of us paradoxically drives us to eat far more than is good for the health of our brain, constantly flooding it with a surfeit of nutrients, night and day.

Humans possess an almost unique emotional connection with our food which makes reducing what we eat difficult. But in terms of brain health, the evidence clearly favours *hara hatchi bu*. Shock the brain!

4

Bugs in the brain

ABOUT SIX THOUSAND YEARS ago on the coast of Lolland, a small island off the Danish peninsula, a young, dark-skinned female sat on the rocks overlooking the ocean and chewed pitch made from heated birch bark. Then she spat it out. This unremarkable occurrence was to give us a remarkable insight into the ancient collection of bacteria that lived in our ancestors' bodies – the hunter-gatherer microbiota. For the first time scientists were able to work out the bacterial content of the early human mouth. The chewed pitch – and the DNA within it – were preserved thanks to the low water content in the pitch, the presence of natural anti-septics and the low oxygen content of the mud in which it became compacted – and fortuitously discovered as archaeologists scrambled to rescue remains in the way of construction to link two islands in the Danish archipelago. The study, published in *Nature* in 2019, together with others examining coprolites (preserved human faeces), revealed that the early human microbiome – the genes in the microbiota – is far removed from our own. Today's urban lifestyle and Western diets have resulted in dramatic changes in the microbes that live in the human gut. It is estimated that 75 per cent of the world's food is produced from only twelve plant and five animal species. We have lost many bacterial 'old friends', with consequences for the health of our body, the health of our gut *and the health of our brain.*

We are not us

There are between 30 trillion and 37 trillion cells in the human body – and, according to the latest research by Israeli scientists, reported by *Nature* in 2016, 50 trillion micro-organisms. That means we are 43 per cent human and 57 per cent microbes – no matter how well or badly we wash! Quite simply, we are not us. We are a 'super-organism', a collective composed of human cells and of microbes which teem over every surface of our body, inside and out. In our skin. In our ears, nose and mouth. In our sexual organs and secretions. In our eyeballs. And – you guessed it – in our gut.

Termed the human microbiota, this population of gut microbes consists largely of bacteria, along with viruses, some fungi and a few organisms of another type called archaea. In 2019, scientists in Cambridge estimated that there are some two thousand bacterial species living in and on us. (You might wonder how we get into the gut or sample from it to identify all these bacteria, many of which can't be grown in the lab. It's very clever. Scientists perform faecal DNA analyses and identify the bacteria from their nucleic acids.) And, as if our cellular inferiority were not enough, consider the genomes. Our cells have a collective genome of some twenty thousand genes. But add up all the genes in our microbiome and you arrive at a staggering number – between 2 million and 20 million. When we consider that each of these is producing thousands of metabolites – biologically active chemicals – we arrive at the inescapable conclusion that *we are as much the product of the bacteria inside us as we are of ourselves.* As our insides gurgle and murmur after a satisfying lunch, a complex chemistry is going on, read by the nerves of the gut and communicated to our brains. We eat badly at our peril. Nervous state, moods, anxieties, fatigue and even our thinking processes are influenced by this heady mix.

A substantial proportion of that heady mix has been there for quite some time, far longer than the chewed piece of pitch has been preserved. We are talking of organisms more ancient, more numerous and more deeply embedded than bacteria. We are talking of viruses, organisms so old that they inserted themselves into our DNA in our early evolution, by infecting sperm or eggs. The development of the human embryo is absolutely dependent on a viral gene which produces a protein called syncytin. Unwittingly, infection by viruses has been a mutually beneficial accident of evolution, an unlikely but necessary form of peaceful coexistence. We perhaps do well to remember this as we face the malignant viral infection of the Covid-19 epidemic.

Given that our gut is largely sterile when we are born, how do the bugs get inside us? In a number of fascinating ways. As the baby is born, the first shot comes from the mother's vagina (or from the skin in a Caesarean), which hosts its own 'garden' or 'microflora'. The second shot comes during feeding, either from a bottle or from the mother's breast, which again has its own internal population of microbes and antibodies, called the 'mobile genetic element'. Mother's milk isn't sterile. But it doesn't usually contain any pathogens. About 25 per cent of gut flora comes from breast milk and about 10 per cent from the skin bacteria growing on the nipple. Thus, right from the start, our microbiota has a high antibiotic component to shelter it from external infection. The third shot comes from all food and anything else which gets into the baby's mouth!

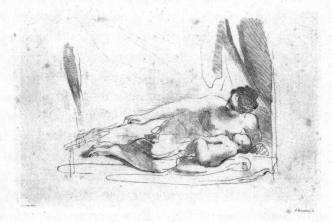

A woman lying down breast-feeding her baby

It takes three years from birth for our individual gut microbiota to become established. It's highly important that during these years the growing immune system of the child remains 'tolerogenic' (generating tolerance) and does not attack the developing bacterial community, in order that the gut microbiota matures properly. This tolerance is achieved to a large degree by special cells throughout the body – called 'dendritic cells' – which modify or downgrade immune responses. These cells were discovered in 1973 by a Canadian physician, Ralph Steinman, who earned a Nobel Prize for his trouble. We therefore 'tolerate' to a degree 'foreign' or invading pathogens and antigens. Once a balanced gut microbiota is established, the dendritic cells, along with other cells in our lymph nodes, pass messages to the immune system not to attack the normalized gut bacteria.

Once our immune system has reached maturity, it needs protecting from us and we need protecting from it. This separation is achieved by the lining of the intestine, which serves a double purpose. The trick is to allow the small-molecule digestive products, such as glucose, to pass through the lining into the bloodstream while preventing opportunistic microbes (and their metabolites) from entering with them. This lining is made up of epithelial cells which have 'tight junctions' between them, plus a layer of mucus which is rich in a protein called mucin that creates a gel coating. If this separating system fails, we get a condition called 'leaky gut', which can allow microbes through the barrier – resulting in massive inflammation. So important is this protection, indeed, that some 70 per cent of our immune system is found in the gut, and the microbiota here complements our own immune system in keeping inflammation and infection to a minimum.

Happy gut, happy body

We give these bugs a home and they work with us to maintain our health. They eat what we eat, get stressed when we do and react to many other aspects of our lives.

Our personal gut flora remains relatively stable once we reach adulthood, but there are big differences between us as individuals. We all have a unique 'bug fingerprint'. Everyone has six main groups of bacteria living in the gut, two of which (Firmicutes and Bacteroidetes) make up 90 per cent of the population. These six groups are carefully balanced and live in harmony with us. Their respective proportions in the population vary from one part of the gut to another – for example, the balance is different in the mouth and in the colon. In the mouth, the microbial population is largely the same around the world. In the colon, it definitely is not. Our health depends on a balance between us and the microbes, and among the microbes themselves. Tiny imbalances make a huge difference. Let us look at one slightly technical example. In a paper published in *Nature* in 2019, Dr David Zeevi of the Weizmann Institute showed that in the DNA of the human microbiome there are a staggering seven thousand structural variations. Each one may be a risk factor for disease – or a valuable addition to our body chemistry. One exotically named bacterium in our gut, *Anaerostipes hadrus*, has a variation in its DNA which ferments inositol (a kind of alcohol) to make a substance called butyrate. This is a ketone and is very important. It not only nourishes the cells of the intestinal lining but reduces the risk of inflammatory disease in the bowel. About 20 per cent of the population have irritable bowel symptoms and no *hadrus* variation.

Many factors can disrupt the balance of our gut flora. They include some you might expect – such as what we eat, our food safety culture, and our dietary and feeding habits – as well as some that might come as more of a surprise, such as how fat we are, how frequently we exercise, how old we are, the season of the year, our lifestyle, gut infections and medicines (especially antibiotics). By way of explanation, ageing alters the composition of our gut microbiota – it becomes less diverse because the immune mechanisms operating in the gut become less effective as we age. It should not surprise us that our gut flora changes seasonally, because the availability of different foods varies considerably by season and the bacterial content varies with the type of food. People also choose to eat different foods in the different seasons. The good news is that many of these factors are under our control, at least to some extent. A full list of the non-dietary factors that can disturb our gut flora is given in box 4.1. If you think the list looks rather long, bear in mind that the first study of its kind, the Flemish Gut Flora Project, found 69 factors – some more important than others – linked to gut flora composition! And that wasn't a 'one-off': the list produced by the Dutch 'LifeLines' project concurred to an impressive degree, with a 90 per cent overlap.

Disturbances in our gut microbe population can have serious effects. They have been linked to obesity, diabetes, cardiovascular disease, liver disease, cancer *and poor brain health*, for example neurodegenerative diseases, which we shall look at later in this chapter. For now, we will look in a bit more detail at a few of the major factors for gut health noted in box 4.1. They are obesity, medicines, exercise, personal relationships and stress.

BOX 4.1: MAJOR NON-DIETARY FACTORS THAT CAN DISTURB THE COMPOSITION OF OUR GUT FLORA

- body mass index (being overweight)
- low levels of habitual physical activity or exercise
- age
- season of the year
- travel across time zones ('jet-lag')
- irregular hours and poor sleep
- excessive psychological stress
- prolonged anxiety
- smoking tobacco
- intimate personal relationships and sexual activity
- medicines and prescription drugs, especially antibiotics
- excessive alcohol consumption

- poor dental hygiene
- dehydration (failure to drink enough water)
- eating erratically, at all times of the day and night
- artificial sweeteners such as aspartame

To begin with obesity, we know that there are major differences between the microbiomes of thin people and obese people: for example, *Bifidobacterium* has been associated with a lean body mass. *Bifidobacterium* is so important that during pregnancy its numbers in the mother's vagina increase dramatically, so that during childbirth the baby ingests a large dose through its mouth. Once in the gut, it digests phytonutrients, produces many essential chemicals (e.g. many vitamins) and is antibiotic, defending the new gut against pathogens. Some species of bacteria show a 'dose response', varying with body mass index or BMI (e.g. *Lactobacillus* accounts for 7, 8, 22 and 34 per cent of the gut population for anorexic, lean, overweight and obese individuals, respectively). And talk to any farmer. Our agricultural friends know from long experience that eliminating large numbers of microbes in the guts of their animals by administering copious antibiotics in their feed makes them gain weight.

The microbiota of the gastrointestinal tract has been shown to be an important factor in the development of obesity. Although it's not yet quite clear how this works, it is well established that modification of the gut microbiota can increase energy production, trigger low-grade inflammation and induce insulin resistance.

Our gut microbial population can also influence our eating and dietary behaviour. For example, some food will taste good because the bacteria need it for their own growth, therefore encouraging us to eat it. Certainly, cravings and unhealthy eating behaviour may be explained by the types of bacteria in the gut. This is a well-known phenomenon in evolutionary biology – bacteria influence the host's behaviour via the brain to increase their own chances of survival – and is called 'behavioural host manipulation'. It has been suggested that one way in which Covid-19 promotes its survival and replication is by infecting the anterior cingulate cortex of the brain. This infection changes our social and emotional behaviour so that when we are pre-symptomatic we are driven to mix with others.

So the advice would appear to be clear: do everything possible to keep your BMI at a healthy level, which means under 25,[1] and keep the gut microbiota in a good balance by improving your dietary intake of micro-organisms (prebiotics and probiotics). We shall look at how to do this later in this chapter.

What about medicines? Many commonly prescribed drugs change our gut microbiota in ways that don't do us any good. Dr Athanasios Typas and his colleagues at the European Molecular Biology Laboratory in Heidelberg tested the effect of 835 non-antibiotic drugs on forty common gut bacteria. Roughly one-quarter of the drugs restrained the growth of at least one bacterium, and nearly 5 per cent affected at least ten. The authors also found that patients taking these non-antibiotics often experienced side-effects similar to those reported for antibiotics. As for antibiotics themselves, a course of a drug such as amoxycillin can decimate our gut bacteria, after which it can take several weeks or even months to restore the normal gut microbiome. The rapidly increasing use of antibiotics since 1944 is thought to have produced widespread changes in our 'global' gut population, including the production of our own internal longlasting reservoirs of antibiotic resistance. The advice appears to be to talk to your physician about the possible effects of any proposed prescription medicines, especially antibiotics.

As for exercise, we've already seen how it improves our overall health and our brain health, so it should come as no surprise to learn that it also improves the health of our gut flora. Studies of people who exercise even moderately show positive changes in gut microbial chemistry, such as increased production of amino acids, carbohydrates and natural antibiotics. A study on women aged 18–40 carried out in 2017 by Dr Carlo Bressa at the European University in Spain showed some of the benefits of physical exercise. The study analysed differences in gut microbiota between women with an active lifestyle (who performed at least ten hours of exercise over a seven-day period) and sedentary women (who performed less than thirty minutes of moderate exercise three times per week). The researchers found significant differences between the two groups in the amounts of eleven types of gut bacteria; they also found, importantly, that the active group had many more health-promoting bacteria. Bressa and his colleagues also found that body fat, muscle mass and activity levels were significantly correlated with several bacterial populations. Basically, leading an active lifestyle improved the gut bacteria balance. Another major study, this time in men, investigated the microbiomes of forty professional rugby union players and compared them to those of non-athletes in control groups. It found significantly greater microbe diversity among the rugby players. The message is clear: regular exercise will contribute to a more diversified and healthier gut microflora.

What could our personal relationships possibly have to do with our gut health, you might ask? Well, almost everyone kisses – whether it's mother and child, spouses or partners, or just friends. Kissing is common in all primates; and research has found that a ten-second kiss can transfer as many as 80 million bacteria into the mouth. And kissing is only one of many behaviours in which we make intimate contact with others. Whether it is sharing food or drinks or utensils, touching, or exchanging breath or bodily fluids, we regularly infect each other from our natural population of microbes. It is a natural part of life – and a highly beneficial one.

There is no better example of these benefits than the report of the Wisconsin Longitudinal Study published in *Nature* in 2019. This study, now in its sixtieth year, has found that social behaviour with family and friends is associated with differences in the human faecal microbiota. Spouses had more similar gut microbiomes than others (e.g. siblings), and

A casual kiss

a better balance, with more diversity. The authors stated that 'married individuals harbor microbial communities of greater diversity and richness relative to those living alone, with the greatest diversity among couples reporting close relationships, which is notable given decades of research documenting the health benefits of marriage'.[2]

It would therefore appear that living in a close relationship with someone is beneficial to the health of our gut microbiota. How close is close? With alarming precision, research has shown that couples who kiss nine times per day have equalized the content of their gut microbes! And the gender of the couples does not matter.

So: exercise is good for us, personal intimacy is good for us – and too much stress, you will not be surprised to learn, isn't. Numerous studies have shown that psychological stress suppresses beneficial bacteria in the gut (and, interestingly, our immune system generally). A 2011 study on mice published in *Brain, Behavior and Immunity* reported that sharing a cage with more aggressive mice suppressed beneficial gut bacteria, decreased their diversity and promoted the growth of harmful bacteria, making the stressed mice more susceptible to infection and causing inflammation in the gut. In a follow-up study, the same investigators found that giving mice antibiotics to reduce the microbiota prevented stress from causing inflammation. However, when the germ-free mice were colonized with a normal population of bacteria, stress again generated inflammation in the gut. It has also been found in other 'germ-free' experiments on mice that the presence of a normal gut flora is essential for memory retention, which is dependent on the amygdala. Similar results have been found in humans. One study in Australia showed that exam nerves in university students suppressed beneficial gut bacteria such as *Lactobacilli*.

Later in this chapter we will look in more detail at the consequences of high stress levels for both the gut and the brain; but for now, it is enough to say that anything we can do which can reduce stress levels will have a beneficial effect on our gut microbiome.

'Let food be thy medicine'

Ever since Hippocrates framed those words over two thousand years ago, humans have sought to optimize their health by the food they eat – with only a limited degree of success. Eighty years after the first formal dietary advice given as a public health measure in the UK was issued in 1941, in the middle of the Second World War, our modern society, with its high-fat, high-sugar diet, is plagued by obesity, diabetes and high blood pressure – indeed, the very dangerous combination of all three, known as 'metabolic syndrome', affects a staggering one in three people over the age of 50 in the UK. For millennia up to and including the time of pitch-chewing hunter-gatherers in Europe, gut microbiomes evolved under conditions of regular exposure to a diverse range of plant and animal microbes that are no longer present in today's processed food chain. Modern foods are very different from that ancient diet, and so are our gut microbiomes. Science is now revealing that *the type of food we eat* affects our health by changing not only the numbers of gut bacteria we host but their diversity, their chemistry and their relationship with the hosting body (us). It's even possible to predict changes in the gut microbes on the basis of what we are about to eat. And, even more astonishingly, the exact same food consumed by two different individuals *can have the opposite effect* in those two people because of differences in their gut microbes. How perverse and counter-intuitive is that? Among other things, it explains why the same diet can cause weight loss in some people but weight gain in others.

Food processing, in a laudable attempt to avoid poor nutrition, has made the nutrients in our food highly 'bio-available' to our small intestine – but in doing so has starved our gut microbiome, most of which is to be found in the large intestine. Processed food covers anything that has been modified to preserve its nutrient content and to add vitamins and minerals, by means of any kind of preserving method. Basically, most of the food that we buy and eat every day is 'processed', from cornflakes to corned beef to curry. One of the blessings of food preservation is that it counters the often short 'half-life' of nutrients in fresh foods – the length of time it takes for the amount in the blood to fall by half – many of which decay fairly quickly after harvesting. But food processing often refines out content with low bio-availability, such as fibre. There's therefore very little food left for our gut bacteria once this type of food has passed through the small intestine. The outcome has been not only increased body weight and obesity but a diminished gut microbiota. Put simply, high bio-availability means fat body and starved gut, whereas low bio-availability means thin body and healthy gut. We need to put our modern food paradigm into reverse, resume the ancient model and thereby restore low bio-availability and with it our old bacterial friends.

How can we do this? The principles of how to achieve a microbe-healthy diet are quite simple and easy to follow:

- The more different types of food we eat, the better. A higher variability of food means a higher diversity of gut microbes and, in turn, lower levels of inflammation and a leaner body. A diverse microbiota produces more neurotransmitters and hormones, reduces pressure on the body's stress system and promotes the maintenance of stable conditions in the brain (homeostasis). Examples of a varied range of foods are shown in box 4.2.

- The food we eat should be less processed and degraded, whether in the factory or in the kitchen. It should have a slowly digestible, natural matrix of fibre (sometimes called a 'prebiotic'). Our Palaeolithic ancestors ate about 100 grams of fibre per day, compared to the 15 grams or so of fibre in a typical Western diet today. Fibre means fewer calories for the host (us), a low bio-availability of nutrients, but high levels of substrate for our gut microbes. This does not necessarily mean that food should be raw. Cooking by steaming or in a microwave preserves more nutrients than other methods. The food is made palatable, but in such a state as to slow down digestion.

- Consuming natural foods which contain high populations of bacteria, called 'probiotics', helps to maintain gut health. Eating raw fruit whole has been found to provide substantial benefits. An apple contains as many as 100 million bacteria and is hugely more diverse (and cheaper) *than any probiotic supplement*. For the evidence, take a look at box 4.3. Probiotic supplements are used to normalize gut microflora in people suffering from diseases such as irritable bowel syndrome (IBS) and obesity, but the evidence for general health benefits, e.g. for the immune system, is equivocal. Not many studies to date have revealed such benefits in healthy individuals. And there is not much evidence that small volumes of supplemental probiotics colonize the gut, though it is not unusual for suppliers of these products to make extravagant claims of this kind.

- A diet rich in plants, vegetables and fruit provides nutrients which are difficult to absorb in the small intestine but pass through to feed the gut microbes. A good example are the substances called 'polyphenols'. These molecules have very low 'bio-availability' – only 1–5 per cent are absorbed in the small intestine. But our gut flora love them and break them down to produce phenolic acids, which are very important for brain function. Phenolic acids are neuroprotective, anti-inflammatory and anti-oxidant, scavenging harmful 'free radicals'. Examples of specific foods that contain polyphenols are given in box 4.2.

BOX 4.2: HOW TO FEED THE BUGS IN YOUR GUT

- Follow the guiding principle: 'low bio-availability = thin body, healthy gut'.

- Vary the food materials in your diet as widely as possible, including meat, fish, poultry, eggs, seafood, dairy and especially botanical sources such as seeds, wholegrain cereals, plants, vegetables and fruits, in order to maximize the diversity of your gut flora.

- Reduce the consumption of food that is processed, whether in the factory or in your kitchen, increasing the slowly digestible, fibrous content of the diet in order to provide maximum food for your gut flora. High fibre is a great guide to choosing low bio-available foods.

- Avoid artificial sweeteners, such as aspartame, which unbalance your gut flora.

- Consume natural sources of probiotic food with a large diversity of microbes, such as whole fruit, in order to support the diversity of your gut flora. Interestingly, beer is a yeast-laden beverage which is associated with a stable and healthy gut flora.

- Include foods in the diet which contain nutrients such as flavonoids (more on these in chapter 5) and polyphenols (found in tea, coffee, dark chocolate, spices, wine, soy, chicory, artichokes, red onions, spinach and red grapes) in order to provide a critical mass of vital nutrients for the gut microflora.

- As a guide, eat more wholefoods, brightly coloured fruits and seasonal vegetables, all of which are rich in fibre and tend to have low bio-availability.

BOX 4.3: APPLES – IT'S ALL IN THE CORE!

- In 2019, scientists at Graz University in Austria examined the humble apple, publishing their results in *Frontiers of Microbiology*.

- They assessed the bacterial content (numbers and types) of all the parts of an apple, including stem, peel, fruit pulp, seeds and calyx.

- They compared organic and conventionally grown apples.

- They found that an average 240g (8oz) apple contained 100 million bacteria, with *90 per cent in the core* and 10 per cent in the pulp.

- Organic apples contained a much more diverse bacterial community – including, for example, *Lactobacillus* and *Methylobacterium*.

- An apple is a natural probiotic – but only if we eat all of it!

It's worth dwelling a little longer on polyphenols. There is a bacterium in our gut called *Akkermansia muciniphila*, named after the microbial ecologist Antoon Akkermans. You may not have heard of it, but this little bug is pretty important. Only discovered in 2004, it munches on the mucus layer of our gut lining and produces nutrients (ketones) which in turn nourish the bacteria that make mucin, a crucial protein discussed earlier in this chapter, so strengthening the gut wall and reducing the risk of microbe penetration. Mice who were overfed so that they were three times as fat as their lean cousins and who were then fed this bacterium *lost half their weight without any change to their diet*. Astonishing. Research in humans has found *Akkermansia* to be abundant in lean individuals and much rarer in people with obesity, IBS and/or type 2 diabetes. It looks like *Akkermansia* is both anti-inflammatory and anti-diabetic – and acts very much like the drug Metformin, used to treat diabetes.

The link between diet and *Akkermansia* was dramatically shown in another study where mice were fed diets different in fat content for a period of eleven weeks. One group was fed lard, while another was given fish oil, and the results were fascinating. In the fish oil group, the abundance of *Akkermansia* increased, as did that of another bacterium, *Lactobacillus*. The opposite happened in the lard-fed mice. Faecal material from both groups was then transplanted into a new group of mice whose original gut microflora had been wiped out by feeding them antibiotics. Once these transplants had been done, all the mice in this new group were fed a diet with lard for fat content for three weeks. The results? In the mice who received the faecal transplants from mice fed on fish oil, *A. muciniphila* levels had risen and levels of inflammation had fallen. In the mice who had received transplants from the lard-fed mice, levels of inflammation were high, and levels of *A. muciniphila* had fallen. And here's the secret: polyphenols are perfect food for *Akkermansia*. The more polyphenols we consume, the better our gut health – and our brain health.

What are we to make of all this? In summary, the general advice must be: take care of the bugs in your gut. We should do everything in our power *not* to disturb our healthy, diversified, well-balanced gut flora. It is essential for general health – and, as we shall see, the health of our brains. Every time we stay up late, travel across time zones, take antibiotics,

Eating the whole apple

drink too much alcohol, smoke or suffer stress, we risk unbalancing our gut flora. And one piece of very bad news: our standardized Western diet does the average gut microbe no favours. As to our diet, we need to understand two things: first, that though we might eat our food in blissful ignorance of what is going on inside us, *our gut microbes largely determine the outcomes of what we eat*; and second, that we need to eat the right foods (summarized in box 4.2) *to look after them*, as well as ourselves.

The bugs in our guts are talking to our brains

We have spent a lot of time in this chapter discussing how to look after the microbes in the gut. That is because their health is vital to the health of our brains – a relationship which has only been discovered in the past ten years or so of research. This is new science.

There is a vast communication highway between the gut and the brain. In chapter 3, I called it the 'gut–brain axis'. Science is now teaching us that the key component in one side of this relationship is our gut microbiota, the 'forgotten organ'. That is, we are talking more accurately about a 'microbiota–brain axis'. Our gut microbes talk to the brain and vice versa, via nerves, hormones and the immune system. In doing so, they greatly influence brain function and behaviour. And when the pathway is disturbed, so is our brain health.

As long ago as 1872, Charles Darwin observed that 'the secretions of the alimentary canal and of certain other organs . . . are affected by strong emotions'.[3] We all know, too, that the brain has a direct effect on our gut. For example, the very thought of eating can release our digestive juices well before our food gets anywhere near the stomach. And we are all familiar with the effects of anxiety – the feeling of nausea, butterflies in the stomach and even loose bowels. We now know that, as mentioned earlier in this chapter, the effects of excessive stress go much further – that it *damages the microbiota itself*. Stress begins in the frontal cortex of the brain with the perception of a threat to our well-being. From there, messages cascade through the brain, down into the limbic system (the seat of our emotions) and into the hypothalamus, the great controller of our basic biology. From here, nervous signals travel directly to the gut via the vagus nerve. These prompt the release of stress hormones such as cortisol and inflammatory molecules such as cytokines; and these in turn cause 'dysbiosis' (unbalancing) of the microbiota. What is more, these effects occur in both directions: that is, a troubled intestine can send signals to the brain, just as a troubled brain can send signals to the gut. Therefore, distress in the stomach or intestines can be the cause *or* the product of anxiety, stress or depression. For example, the stress-related 'cytokine storm' has been shown to disrupt brain neurochemistry and make us more vulnerable to mood changes. A cytokine storm is the sudden uncontrolled release of inflammatory molecules from the immune system, often provoked by virulent infections and characterized by multiple organ failure and high temperature. And more than half of people with chronic gastrointestinal disorders such as Crohn's disease, IBS and other gut-related conditions are plagued by anxiety and depression.

But the reach of our microbes goes way beyond the emotions related to stress, or to the gut itself. Gut microbes are now thought to regulate our thinking and behaviour. If you think that's a crazy idea, you're not alone. The earliest scientific report on the subject emerged in 1998, when Dr Mark Lyte, then at Minneapolis University, showed that introducing a pathological bacterium into the gut caused a change in behaviour. He and his colleagues fed a small dose of the pathogenic (disease-producing) bacterium *Campylobacter jejuni* to a group of lab mice. Two days later, the mice became much more cautious about entering open areas of a laboratory maze – a sure sign of anxiety – compared with mice in a control group. And yes, Lyte was dismissed by the scientific community as crazy. But ten years on, the evidence supporting his thesis had reached a critical mass – and it was evidence from multiple different laboratories, always a good sign in science.

While Lyte showed that harmful bacteria can induce anxiety, others showed (again, in the face of pronounced scepticism at the time) that beneficial bacteria can calm down anxiety-prone mice. In a 2011 study, mice were fed doses of *Lactobacillus* in their food. After twenty-eight days they were faced with the open spaces of a maze and a swim test. Compared to a control group, the *Lactobacillus* mice showed much less anxiety – they entered the maze more freely and showed calm floating behaviour in the 'forced swim' test. Interestingly, later analysis showed that the *Lactobacillus*-fed mice produced lower levels of the stress hormone corticosterone. Furthermore, some brain regions in all the mice showed an increase in the number of receptors for gamma-aminobutyric acid (GABA) – a neurotransmitter that mutes neuronal activity, keeping anxiety in check.

There are many more such examples. Now the world of science has come round to an even more revolutionary idea – that our gut micro-biota influences normal mental processes in humans and is involved in numerous mental disorders and neurological conditions. *And further, that microbes themselves are invading the brain.*

One striking piece of evidence of an association between our gut microflora and our behaviour was produced in 2017 in California by a team led by gastroenterologist Dr Kirsten Tillisch. The researchers took faecal samples from a group of forty healthy women aged between 18 and 55. On the basis of the bacteria found in their microbiomes, the women were divided into two groups: one group of thirty-three was rich in Bacteroides and the other, a smaller group of seven, rich in Prevotella. Next, MRI scans were taken of their brains as they viewed images of individuals, activities or things that evoked emotional responses.

What the researchers found was that the Bacteroides group showed greater thickness of the grey matter in the frontal cortex and in the insula, and a bigger hippocampus – all brain regions involved with complex information processing and memory. By contrast, the Prevotella group showed more connections between emotional, attentional and sensory brain regions, and lower brain volumes in several regions, including the hippocampus. In this group, the hippocampus was less active while the women were looking at negative images, and they also recorded higher levels of negative feelings such as anxiety, distress and irritability. This may be because the hippocampus helps us to regulate our emotions, and so if it is smaller – which may possibly be related to the makeup of our gut microbiota – negative images may pack a greater emotional punch.

Though the group sizes were small, these unusual results support the concept of interactions between the gut microbiota and the brain in healthy humans. This early study set the ball rolling for more research. Two other pieces of evidence are worth noting, especially for the sceptics among you who just can't bring yourselves to believe that the contents of your colon influence your thinking and reasoning.

The first comes from another US study from 2017, looking at cognitive test performance and gut bacteria in a small sample of forty-three healthy people aged between 50 and 85. The group was divided into two on the basis of their test scores. The results showed a clear and significant relationship between the level of performance in the cognitive tests and four of the major bacterial groups in the gut. How might we explain such a finding? There are a number of explanations. One is the known relationship between gut dysbiosis and the level of inflammation transmitted from the gut to the brain – chronic inflammation is a known risk factor for declining cognition. Others include direct neural communication (nerve signals) to the brain via the vagus nerve and circulating metabolites from the gut bacteria.

The second concerns the production of serotonin, an important neuro-transmitter in the brain that has been the focus of many clinical studies. Strikingly, most serotonin is *not produced by the brain itself.* As noted in chapter 3, most of it is the result of a complex chemical cooperation between bacteria in our gut and some of the cells that line the intestine (chromaffin cells). Serotonin receptors are found in the human brain in areas that are important to learning and memory. Studies have found that alterations in serotonin activity influence cognitive performance. The current evidence suggests that reduced neurotransmission by serotonin has a negative influence on cognitive functions, and that normalization of serotonin activity may have beneficial effects.

Across the animal kingdom, in many species, it is known that the presence of micro-organisms in the host can influence social behaviour – how individuals react and relate to each other, and how they communicate. It's known that the microbiome in these cases can communicate with the brain (for example, the amygdala) in three ways: via the immune system ('immune signalling'); via hormones such as serotonin; and via nerve signals through the vagus nerve. Again, these observations shouldn't come as a surprise. The bacteria are changing our social behaviour in order to promote their own survival. But now we also have evidence that treating people with conditions that cause social behaviour deficits, such as autism, by supplementing their diet with certain types of bacteria, lowers levels of anxiety and anti-social behaviour, and improves sociability, speech and language communication.

The main points of what we currently know about the bugs in our gut and the brain are summarized in box 4.4.

BOX 4.4: BUGS AND THE BRAIN: SOME OF WHAT WE KNOW

- A healthy, balanced and diverse gut microbe population helps us to cope better with stress.

- Stress responses disturb the balance of our gut microbe population and these effects feed back into the brain in structures such as the amygdala, part of the limbic system (seat of the emotions).

- Our gut microbe population can influence our dietary behaviour; for example, some food will taste good because the bacteria need it for their own growth, therefore encouraging us to eat it.

- Mood, emotion and even personality are strongly influenced by our gut microbes, which produce neurotransmitters such as serotonin, GABA and dopamine.

- Cognitive functions, including learning capacity and memory, are closely related to the microbes in the gut.

- A normal gut microbiota has even been found essential for the development of social behaviour -- and social behaviour helps the microbes to colonize.

Bugs and brain health

Just as Mark Lyte's ideas about the connections between gut microbes, thinking and behaviour were initially dismissed as crazy, so researchers looking at similar potential links with brain disease fared little better. There is no worse feeling for a scientist than seeing their new research findings receiving an indifferent or even hostile response. And that is exactly what happened in 2014, when Professor John Cryan, a microbiologist from University College Cork in Ireland who had made his academic name in America, parachuted himself into a neuroscience conference in San Diego and tried to tell the audience that gut microbes played a role in the development of Alzheimer's disease. He was met with laughter. As the saying goes, a prophet is not without honour, except in his own country. Well, not for three years, anyway – until researchers at the Wisconsin School of Medicine got hold of the idea. In an elegantly simple procedure, they showed that the gut microbiome of people with Alzheimer's was very different from that of people of the same age and sex not afflicted by the disease – and that these differences were related to biomarkers for Alzheimer's in the fluid around the brain. However, much remained to be ascertained – including the direction of cause and effect: was the disease causing the microbiome to change, or was the microbiome implicated in causing the disease? Other researchers came up with similar results. One of the San Diego sceptics, Dr Sangram Sisodia, took Cryan's ideas away with him to Chicago and decided to test them. He took some mice who were predisposed to Alzheimer's and removed their gut microbes by feeding them antibiotics in their diet. In the brain with Alzheimer's, it's common to find clumps of a protein called beta amyloid, which is very toxic and forms plaques between the neurons. Amazingly, he found that in his treated mice, these clumps of protein were *reduced*. But again, this just led to more questions – in particular, which bacteria were causing the effect, and how? Research on these points was to lead in a surprising direction.

UNFAIR ADVANTAGE.

A visit to the dentist

Do you have bad oral health? Do you resist going to the dentist? Do you have the signs of gingivitis (gum disease)? If any of these are true, *you are risking the health of your brain*. Because the mouth is a primary entry point into the body, poor oral hygiene can spell big problems for your health. Many, many studies have found strong correlations between inflammation in the mouth and cardiovascular disease, respiratory infections, diabetes, kidney disease, cancer and even erectile dysfunction in men. And now we can add brain disease to the list.

The mouth offers a warm, wet, cosy environment which bugs love; so it's not surprising that it's home to an amazing 6 billion bacteria of seven hundred species. They have been in the collective mouth of humanity for millions of years. They talk to each other, modifying their activities by a process called 'quorum sensing', creating a balanced population which confers mutual benefits between us and them (they don't bother us and we don't bother them: a kind of biological peaceful coexistence). But there are also unwelcome visitors. Many of them come from eating. Others from kissing. As we have seen, the more people (or pets) we kiss on the mouth, the more bacteria enter. Not all of them stay and colonize. But some do; and some of them are harmful foreign bacteria that inflame the tissues of the mouth. Science is now finding that the inflammation products *and the bugs themselves* can enter the bloodstream and travel to the brain, sometimes diffusing there through the nerve fibres supplying it. They have been found to kill brain cells and lead to memory loss. The trail of evidence is leading straight to Alzheimer's disease – and to one bug in particular.

Alzheimer's disease bears all the hallmarks of brain infection; but ever since the identification of beta amyloid in 1984, scientists have concentrated on the 'protein accumulation' hypothesis, spending billions worldwide on looking at how two particular proteins, tau and beta amyloid, might cause the disease – sadly, without much success: some 90 per cent of all drug trials have failed. Now scientists are looking in a different direction – at bacteria. Doctors have known for some time that a particular bacterium called *Porphyromonas gingivalis* (the cause of gum disease) was common in patients with Alzheimer's and was a risk factor for the disease. In 2014, researchers in the UK, following up on the work of Dr John Cryan, carried out an unusual experiment using mice. They infected mice mouths with four types of bacterium, including *P. gingivalis*. The results took the world of neuroscience by storm: the bacterium's DNA *ended up in the brains of the mice*. And what about humans? In 2019, researchers from several universities reported finding the two toxic enzymes that *P. gingivalis* uses to feed on human tissue in *over 90 per cent* of fifty-four human Alzheimer's brain samples taken from the hippocampus – a brain area important for memory. These protein-degrading enzymes are called gingipains, and they were found in higher levels in brain tissue that also had more tau fragments and more of the cognitive decline associated with this protein. The team also found DNA from *P. gingivalis* in the cerebral cortex – a region involved in conceptual thinking – in all three Alzheimer's brains in which they looked.

P. gingivalis is not a part of what dentists call a healthy 'climax community' of bugs in our mouths. This black-pigmented, pathogenic bacterium is found lurking throughout the body – not only in the gut but in the respiratory and reproductive systems – and is normally suppressed by the immune system and the presence of a balanced 'bug' community. Any disturbance in either gives *P. gingivalis* the opportunity to strike.

Chew for a bigger, better brain

What an unlikely story! Poor oral hygiene is related not just to tooth decay, more fillings and higher dental bills, but to a whole lot of other undesirable outcomes, including increased risk of a heart attack, erectile dysfunction and *damage to the brain*.

Fortunately, dental health companies have invested a fortune in R&D on our oral hygiene. But there is one message you will never see in their health advice, and most people would stare in disbelief if they heard it: *if you want to look after your brain, chew gum*. How on earth could that work, you may ask? The truthful answer is, we don't really know. But the evidence is so compelling that national governments have used it in their health promotion campaigns.

Drs Lucy Wilkinson and Andrew Scholey at the University of Northumbria studied the positive effects of chewing gum on mental performance. They split a group of seventy-five adults into three groups: one chewed sugar-free gum, a second mimicked chewing without gum (sham chewing) and a third did nothing. Each group undertook tests of attention and working memory. The result: gum chewers showed an improvement in immediate and long-term memory. Their word recall was 35 per cent better than the two other groups.

Now let's look at chewing gum and *brain size*. Dentists in Taiwan compared forty people over age 65 with either mild cognitive impairment or Alzheimer's to a control group of thirty healthy people in the same age group. The participants were given brain scans and had their teeth and chewing assessed. Among the controls, increased chewing ability was linked to larger grey matter volume in the pre-motor cortex of the brain, which helps to control muscle movement. In the group with Alzheimer's or mild cognitive impairment, those with the poorest chewing ability had less total grey matter volume in the brain. Other key brain areas important for memory were also smaller.

If you think these were maverick studies, think again: a review of twenty-three studies conducted across the world found that twenty linked poorer chewing to lower cognitive function; and in eight of them, poorer chewing was a risk factor for mild cognitive impairment or dementia.

The enhancement of memory while chewing gum is what scientists call a 'robust' finding,[4] although the processes underlying the effect are not known – scientists are still working on it. But we can already offer some explanations. Chewing gum appears to serve a double purpose: by increasing the flow of saliva, it makes the mouth less acid; it reduces pathogenic bacteria in the mouth (a risk factor for brain health); *and* it stimulates areas of the brain that are important to thinking and memory. One explanation is that chewing gum prompts the release of insulin, which we know has receptors in the brain. Another is that chewing gum increases our heart rate, so providing the brain with more oxygen and nutrients. Either way, it has a positive effect on our memory.

So, we have a subtle message: millions of helpful microbes in the mouth have to be preserved while pathogenic ones have to be eradicated. How are we to do this? The crucial factor is eliminating the acid-producing and acid-loving bacteria which build up into a sticky biofilm around the teeth that becomes plaque. Biofilms in the mouth are exceptionally nasty. They are resistant to all types of antibiotics and antimicrobial agents, and usually are able to resist the body's own immune system. So an important principle for a healthy mouth is to keep it from becoming unduly acid – or, as scientists would say, to avoid its pH value falling below 5.5 (pH less than 7 = acid; pH 7 or above = alkaline). All dental hygiene methods, including when you should clean your teeth, are based on this principle. A summary of the positive oral care measures you can take is given in box 4.5. The important point is that these measures, taken together, will maintain a healthy oral microbiome and at the same time will reduce widespread inflammation and the risks of disease elsewhere in the body, including poor brain health. In summary, it is a question of helping the 'good' bacteria in the mouth and suppressing the 'bad' bacteria. If our mouths are acid or our teeth aren't clean, bad bacteria will thrive.

Microbes and the mind

In a paper in *Nature* in 2012 entitled 'Mind-altering micro-organisms', two of the world's leading experts remarked: 'A growing body of evidence indicates that microbiota have a role in the normal regulation of behaviour and brain chemistry that are relevant to mood.'[5] Moreover, they intriguingly claimed that our microbiota composition may influence our susceptibility to two common modern-day afflictions: anxiety and depression. A burgeoning body of evidence since then has shown that our nineteenth-century forebears who practised colonic irrigation, purging and even bowel surgery may not have been that far wide of the mark in their focus on this area. The vast assemblage of microbes in our gut does, it appears, have a big part in framing our state of mind. And that includes both anxiety and depression.

BOX 4.5: TAKING CARE OF YOUR MOUTH – PROTECTING THE GOOD BUGS, ELIMINATING THE BAD ONES

- Use anti-biofilm mouthwash (containing chlorhexidine gluconate) sparingly.
- Clean teeth twice a day, counting five seconds to clean each tooth.
- Twice a day means before breakfast and before retiring to bed.
- Floss once every day.
- Rinse mouth with non-sparkling water after eating.
- Try not to eat between meals.
- Do not smoke.
- Drink water regularly throughout the day.
- Reduce consumption of sugary foods and drink – and acidic low-sugar drinks.
- Maintain a healthy diet.
- Do not drink alcohol excessively, particularly spirits, and avoid binge drinking.
- Avoid the use of antibiotics unless absolutely essential.
- Visit your dentist regularly.
- Avoid getting oral piercings.
- Chew sugar-free gum!

The key to unravelling these issues lay with germ-free mice – simply because we can carry out experiments on them that we could not do on humans. A research team at McMaster University in Ontario discovered that if they colonized the intestines of germ-free mice with bacteria taken from normal mice, the recipient mice would take on *aspects of the donors' personality*. Naturally timid mice 'came out of their shells', whereas more daring mice became 'shrinking violets'. These findings suggested that inter-actions between bugs and the brain could change personality and mood.

Now to humans. It is noticeable that people who suffer from gut illness often also have anxiety and depression that cannot be fully explained as emotional side-effects of their illnesses. And more than 50 per cent of people with IBS also have depression and anxiety alongside that affliction. How might we explain these observations?

Several studies have shown a two-way link between depression and the gut microbiome. We know that depression is associated with an imbalance in the link between the hypothalamus, the pituitary gland in the brain and the hormone system (the hypothalamic–pituitary–adrenal or HPA axis), which causes the release of cytokines from the immune cells of the gut. These molecules circulate in the blood and in turn trigger the release of cortisol, a potent stress hormone, from the adrenal gland. Cortisol contributes to anxiety and depression. Conversely, it's been observed that improvements in the symptoms of depression often restore normal HPA activity. It appears, then, that the action of our gut and its unique bacterial community is central to maintaining mood stability. Yet again, we see that looking after our microbes is far from a trivial matter.

But the really amazing linkage was the one found between autism and gut bacteria by the late Paul Patterson, a neuroscientist at the California Institute of Technology. He noticed that women who suffered from a high fever during pregnancy were up to seven times more likely to have a child with autism. This did not look like genetics! So he induced flu-like symptoms in pregnant mice using a chemical which was a 'viral mimic'. Amazingly, the offspring of Patterson's mice displayed all three of the core features of human autism: limited social interaction, repeti-tive behaviour and reduced communication. Moreover, the mice had 'leaky gut', which was important because between 40 and 90 per cent of all autistic children also suffer from gastrointestinal symptoms of this kind. Now, no one is claiming a 'microbial cure' for autism. But these new findings do offer vital clues that suggest gut microbes contribute to the condition.

Finally, there's Parkinson's disease – a slowly progressive disorder that affects balance, movement and muscle control. It is the second most common age-related neurodegenerative disorder after Alzheimer's, and currently affects about 145,000 people in the UK. It used to be considered purely a 'brain disease', but two inconvenient facts have challenged this view: first, as many as 90 per cent of cases are idiopathic (that is, without any identifiable cause); and second – amazingly – changes in the nerve cells of the gut are found years before specific Parkinson's symptoms appear. Then, in 2014, came an unexpected breakthrough: a study in Helsinki found that the intestinal microbiome is altered in people with Parkinson's, and that this change is related to the disease's effects on movement. Unsurprisingly, this finding prompted a welter of follow-up studies. All have replicated the finding that Parkinson's patients show changes in the intestinal microbiome, leading to a major change in thinking about this disease: it is now believed that the (as yet unknown) environmental factors that trigger Parkinson's *do so via the gut microbiome – and not via one single bacterium but by disturbing the equilibrium of the whole flora.* So a game-changing hypothesis has been formulated: does Parkinson's disease start in the microbiome of the gut and spread to the brain via the vagus nerve?

A list of the neurological diseases we now know to be associated with changes in the gut flora is given in box 4.6.

BOX 4.6: BRAIN DISORDERS KNOWN TO BE ASSOCIATED WITH CHANGES IN THE GUT MICROBIOTA

- Alzheimer's disease
- anxiety
- depression
- autism
- Parkinson's disease

The physician in your gut

Deep within us, in the dark, cavernous, airless reaches of our gut, is an amazing factory. Churning silently away, it talks to the gut – and to the brain. It reads the chemistry of our body and reacts accordingly, using recipes tested over millions of years, to dispense a formulary of medicinal compounds. They maintain our health; they affect our moods and emotions, and even our thinking skills. Key advice on how to protect this dispensary and thereby your brain health is summarized in box 4.7.

BOX 4.7: TAKE CARE OF YOUR GUT, TAKE CARE OF YOUR BRAIN

- Keep at a sensible weight, with a BMI of less than 25.
- Monitor carefully what medicines you are taking, especially antibiotics.
- Exercise according to official recommendations.
- Invest in a close, stable, personal relationship.
- Do everything to reduce harmful psychological stress.
- Pay attention to lifestyle factors which cause dysbiosis.
- Eat a high-fibre, low bio-availability diet and natural probiotic foods.
- Maintain good oral hygiene.
- Chew gum!
- Don't worry too much about dirty surfaces. We live in an ocean of microbes, most of which are beneficial; and without the harmful ones, we would have no immunity.

In the fourth century, the Chinese physician Ge Hong prescribed a treatment called 'Yellow Broth'. It was administered to patients by mouth.

Shockingly, its main component was human faeces from a healthy person. Today, the idea seems totally disgusting to us – but it's worth recalling that the annals of medicine are replete with examples of the practice, right up to the present day. Today's physicians call it 'bacterial therapy' or 'faecal microbiota transplantation' (FMT). It is administered by oral capsule or by colonoscopy, in treatment for diseases of the gut involving dysbiosis. The intriguing question is: can we manipulate our gut bacteria to treat or improve the health of the brain? FMT is just one method currently used to manipulate our gut flora. Others include probiotics, prebiotics, personalized diet and prescription lifestyle changes. In the future, it is possible that we will be able to harness our gut microbiota to improve brain and mental health and to prevent and treat diseases of the brain. If this seems outlandish, just consider what we readily accept at the moment: psychotropic drug regimes with sobering side-effects, invasive brain surgery and even gene manipulation (via 'transposons' – mobile segments of DNA). How enticing it is that the maintenance of our brain health, and even radically reducing the risk of Alzheimer's, may rest in something as simple as good oral hygiene; or that a simple faecal transplant from friends or family might be the future cure for many brain disorders?

Recent cutting-edge research has given birth to a new range of medical treatments called 'psychobiotics'. A psychobiotic is any substance that can modify the signals that pass between microbes, gut and brain to produce a psychological effect. They can be anti-depressant and anxiety-reducing. Sounds fanciful? In the past ten years, over two hundred microbiome companies have been set up, prompting Forbes to call it 'the decade of the microbiome'. There are currently some 650 research programmes in progress, forty of them focused on the gut–brain axis, developing strategies to treat autism, Parkinson's, Alzheimer's and depression. And the inevitable marketplace has emerged, with new products on offer at pharmacies and in convenience stores and through countless online retailers. It's a brave new world for those who wish to try it.

Food for thought

OVER A HUNDRED YEARS ago, a team of scientists at Cambridge University led by the remarkable Professor Frederick Gowland Hopkins made a milestone breakthrough in nutritional science. They discovered vitamins. Hopkins and his colleagues were astounded to find that the growth of young piglets could be halted even though they were being fed ideal quantities of pure macronutrients – carbohydrates, proteins and fats – plus adequate minerals (micronutrients) and water. Why should this be? Hopkins hypothesized the existence of 'accessory growth factors' – later identified as 'vitamins'. For this work he was awarded the Nobel Prize in Physiology or Medicine in 1929. And has deserved the gratitude of humanity ever since.

But there is a big problem with vitamins. Our bodies can't make most of them. We – and that includes our brains – are completely at the mercy of our diet. If we wish to optimize our brain health, ensure correct brain development in our children or slow down the ageing of our own brains, the correct intake of vitamins is essential.

Great skin and a better memory

On 28 March 2007, the BBC broadcast a documentary on the cosmetics industry in which they interviewed Dr Chris Griffiths, a dermatologist at Manchester University. He casually mentioned that there was a commercial women's skin cream on the market, Boots No. 7, that was able to reverse sun-inflicted skin damage. The next day sales rocketed by 2,000 per cent. Women cleared the shelves and the company's webstore ran out. Rationing was introduced. The cream contained Vitamin A.

But Vitamin A will not just give you great skin. It will also give you a great brain. It was in 1998 that researchers at the Salk Institute for Biological Studies, San Diego, not only discovered that Vitamin A promotes learning but found the region of the brain where the Vitamin A receptors are located. It was the hippocampus, where this particular vitamin stimulates the synapses involved in learning. Currently, 190 million children worldwide are thought to be deficient in Vitamin A.

Furthermore, it's also been found that Vitamin A contributes to the neurobiological processes that underlie memory performance *throughout life* – and that it *prevents, limits or delays age-related cognitive decline*. What's really interesting is that it's not simply inadequate intake that contributes to age-related change but the way in which Vitamin A's signalling chemistry changes in the brain as we get older. That has not stopped people asking the question: can we maintain youthful brain function by supplementing our Vitamin A intake?

Researchers at the University of Bordeaux in France looked at the effects of Vitamin A deprivation and subsequent reintroduction in rats. They found that Vitamin A-deprived rats produced 32 per cent fewer new cells in the hippocampus than those fed adequate amounts of Vitamin A. The deprived rats also did much less well than the control group in a practical test – finding an underwater platform in a swim test – taking as much as 25 per cent longer to succeed. The clincher was, of course, what happened when Vitamin A was added back into the diets of the deprived rats. After four weeks of reintroduction – Bingo! They out-performed a control group of 'normal' diet animals. Not only did their performance improve; when examined, the rate at which new neurons were being made exceeded their normal rate, as if they were compensating for the Vitamin A 'break'.

So how much Vitamin A should we be getting? In the UK, the government recommendations for adults (age 19 and over) are 600 micrograms (mcg) a day for women and 700mcg a day for men (1mcg = one-thousandth of a milligram, so one-millionth of a gram). Food manufacturers, meanwhile, often measure vitamin content in 'international units' (IUs), 1IU being equal to 0.3mcg of retinol (the chemical name for the main form of Vitamin A). Therefore, to convert your recommended daily intake into IUs, you just need to divide the number of micrograms by 0.3; for example, 600mcg translates into 2,000IUs.

For an idea of how much Vitamin A you're getting, take a look at box 5.1, which lists the amounts in common sources of the vitamin, per typical serving. Not shown in the table are values for fish oil, consumed 'neat'. Fish oils are one of the richest natural sources of Vitamin A – for example, just one teaspoon of cod liver oil contains 1,350mcg, over twice the recommended daily intake for women and nearly twice that for men.

BOX 5.1: FOOD SOURCES OF VITAMIN A

Food	Typical serving	Vitamin A (mcg)
Ox liver	3oz (85g)	6,582
Sweet potato (baked)	Medium (100g)	960
Spinach, boiled	4oz (115g)	573
Carrots, raw, grated	0.9oz (25g)	459
Ice cream, vanilla	5oz (150g)	278
Egg, boiled	One, large	75
Broccoli, boiled	3oz (85g)	60
Salmon, poached	3oz (85g)	59
Tuna (in oil, drained)	3oz (85g)	20
Chicken, roasted	4oz (100g)	5

Source: US Department of Agriculture Food Composition Database.

B is for brain

The B vitamins (B1, B2, B3, B6 and B12) all help to maintain a healthy nervous system. They literally 'make things happen' in the brain. They play a critical role regulating energy release in brain cells and enabling the action of neurotransmitters, the chemical messengers that transfer information between our 86 billion neurons. All of our movements, thoughts and feelings depend on our neurons talking to each other via these messenger molecules. Neurotransmitter imbalances can cause problems such as fatigue, confusion, anxiety, depression and hormonal dysfunction. In this section of the chapter I will concentrate on the B vitamins for which the evidence of neurotransmitter support is strongest – choline, Vitamin B6 and Vitamin B12.

Choline is probably one of those 'brain foods' of which you've never heard. Although not technically a vitamin by definition, this neglected

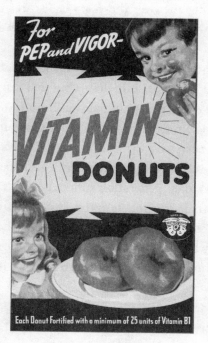

Getting your vitamins

substance is probably one of the most important of additional nutrients. Described by the US National Institutes of Health as 'an essential nutrient for public health', it is a vital ingredient in the diet of children, adults and older people alike. Yet for nearly fifty years after Steven Zeisel discovered in 1977 that it was an essential component of breast milk, medical orthodoxy casually ignored the evidence of its significance. And its value as a brain nutrient wasn't recognized until 2011.

Choline is a primary building block of the key neurotransmitter acetylcholine; it is a vital ingredient of the membranes of brain cells; and it enables the transmission of signals across the synapses – the gaps between neurons – that is essential for learning, memory, mental clarity, focus and concentration. And the proof of its effects? In 2011 a dietary study was carried out in which 1,391 young people took tests of verbal memory, learning and reasoning. Their test performance was directly related to their choline intake, after accounting for all other likely factors. Cognitive decline over the years is also related to the choline in our diet. Choline is thought to protect the brain from Alzheimer's disease in two ways – first, by reducing levels of homocysteine, an amino acid that doubles the risk of Alzheimer's; and second, by 'calming down' microglia, cells that clear out brain debris but that can get out of control, contributing to the development of the condition.

On top of all this, an experiment by researchers in Arizona in 2019 showed that choline can have a transgenerational effect. Such generation-leaping effects were already known to science – for example, children of the 1944 'Hunger Winter' generation suffered from 'silenced genes' as a result of their parents' starvation, resulting in a high prevalence of obesity among their own children in the post-war years. The 2019 experiment showed that mice which received supplemental choline in their diet had offspring which showed significant improvements in certain types of memory.

Fortunately for us, choline is abundant in all types of food, though at considerably lower levels in plant food sources. It's hard to quantify your individual need for choline because of genetic variation within the human population, which current recommendations do not take into consideration. So the guidelines are set in terms of 'adequate intake' rather than a recommended daily allowance or RDA. In the EU, adequate intakes are set at 425 milligrams (mg) a day for women, 450mg a day for pregnant women, 550mg a day for lactating women and 550mg a day for men.

Having said all this, dietary choline is an excellent example of how difficult it is to say precisely *how* a particular nutrient is related to health. Many authorities, including the renowned Harvard Health Publishing (a consumer health education body attached to the university's medical school), have warned against eating too much choline-rich food because it raises the levels of a recently discovered molecule called TMAO, which plays a part in heart disease. Now there is evidence from Europe showing that a fish-rich and vegetarian diet, which is beneficial or at least neutral for cardiovascular risk, is associated with a *significantly higher level of TMAO in blood plasma than diets rich in red meat and eggs.* What are we to do? The answer is essentially very simple: eat a balanced diet, stay within the guidelines and do not take choline supplements. You can work out what you are getting by looking at box 5.2, which sets out a range of foods that offer abundant supplies of choline.

BOX 5.2: FOOD SOURCES OF CHOLINE

Food	Typical serving	Choline (mg)
Ox liver	3oz (85g)	356
Wheatgerm, toasted	6oz (150g)	202
Egg	1 large	147
Beef, roast	3oz (85g)	97
Scallop, steamed	3oz (85g)	94
Salmon, canned	3oz (85g)	75
Chicken, roast	3oz (85g)	73

Atlantic cod, baked	3oz (85g)	71
Broccoli, boiled	3oz (85g)	63
Peanut butter (smooth)	0.5oz (14g)	20

Source: US Department of Agriculture Food Composition Database.

On 31 July 2016, the most dangerous stunt ever attempted was shown live on US television. Luke Aikins, a 42-year-old skydiver, leaped out of an aircraft at 25,000 feet *without a parachute*. After a terrifying two-minute free fall, he landed just off dead centre in a 100-foot by 100-foot net at the Big Sky movie ranch on the outskirts of Simi Valley, got up and walked away. What, you are well entitled to ask, has this death-defying adventure got to do with nutrition and brain health? It is this: analysis of Luke Aikins' blood after the event by expert nutritionist and event organizer Chris Talley revealed that the skydiver's system had been completely stripped of all traces of Vitamin B6. *Two minutes of acute stress, and the weeks leading up to it, had wiped out his body's store of this key vitamin.* There is a huge message here for people who are going through a divorce, dealing with a dying relative or trying to prevent their business from going bankrupt. Vitamin B6 is indispensable to stress resistance. No wonder Luke's was zeroed out.

Nor is it any surprise that the media call B6 nature's 'anti-stress' vitamin. Research has shown that B6 (or rather, its active form, P5P) is involved in the production of major neurotransmitters in the brain, notably serotonin, dopamine, adrenaline, noradrenaline and GABA – all of which have considerable impact not just on cognitive development but on our mood states as well, including levels of depression and anxiety. It also plays a part in keeping down blood pressure and reducing the impact of the stress hormones corticosteroids. It appears that B6 works together with magnesium to reduce stress, to the point that the pharmaceutical company Sanofi has a combined B6–magnesium product in a Phase IV clinical trial. All 264 participants in the Sanofi study showed stress re-duction after eight weeks of taking magnesium alone or a combination of magnesium with Vitamin B6. But those taking the combination, regardless of their stress level (severe or very severe), showed a 24 per cent stronger and significant positive effect.

Quite simply, mental toughness and endurance rely heavily on an adequate intake of Vitamin B6. So how much do we need, and how do we get it from our diet? The UK government recommends 1.2mg a day for women and 1.4mg a day for men. In the USA, the recommendation is 1.3mg per day for both men and women, rising to 1.7mg a day for men and 1.5mg a day for women after age 50 (they don't say why). The richest sources of Vitamin B6 are fish, liver and other offal, potatoes and other starchy vegetables, and fruit (except citrus). You can look at what you are getting by checking out box 5.3.

BOX 5.3: FOOD SOURCES OF VITAMIN B6

Food	Typical serving	Vitamin B6 (mg)
Chickpeas (canned)	1 cup (8oz)	1.1
Ox liver	3oz (85g)	0.9
Tuna, cooked	3oz (85g)	0.9
Salmon, poached	3oz (85g)	0.6
Chicken breast, roast	3oz (85g)	0.5
Potatoes, boiled	1 cup (8oz)	0.4
Banana	1 medium	0.4
Beef patty	3oz (85g)	0.3
Cottage cheese	1 cup (8oz)	0.2
White rice, boiled	1 cup (8oz)	0.1
Raisins	½ cup (4oz)	0.1

Source: US Department of Agriculture Food Composition Database.

Low levels of Vitamin B12 in the body are common in Western societies – which is not good news, because this vitamin is essential for normal health and one of the most important vitamins for brain health! Vitamin B12 acts as a 'co-enzyme' in our cells: that is, it stimulates essential chemical reactions such as the production of DNA and the manufacture of myelin, the white fatty sheath around our nerve fibres. It is thus essential to processing speed. Paradoxically, one of the principal reasons underlying the high prevalence of Vitamin B12 deficiency is a health-related dietary choice – a vegetarian or vegan diet. Scientific studies have shown that up to 80 per cent of vegans are deficient in Vitamin B12. To illustrate the consequences of deficiency in this crucial nutrient, consider the following case study of an otherwise healthy patient from the St Joseph's University Hospital, Chicago: 'A 52-year-old male was initially presented in an ambulatory clinic complaining of fatigue and weakness for 2 weeks. His fatigue was accompanied by dyspnea [shortness of breath] on exertion and light-headedness which have increased in frequency in the last 4–5 days prior to the presentation. Physically, he looked thin and pale.'[1] Even small deficiencies of B12 lead to fatigue, brain 'fogginess' and negative mood, which can tip into depression. Chronic deficiency is a known cause of some dementias and may even cause permanent damage to the brain. Symptoms of deficiency include problems with short-term memory, recognizing people and places, finding the right words, problem-solving, planning and carrying out simple tasks, exercising judgement, and controlling moods and behaviour.

The UK government RDA of Vitamin B12 for adults aged 19 and over is 1.5mcg. This is a minimum figure; other authorities recommend more. In the USA and Canada, the guideline is 2.4mcg a day, and in Europe it's 4mcg a day. Readers should remember that 1mcg (microgram) is one-thousandth part of 1mg (milligram). The major sources of B12 are all animal – eggs, meat, liver, fish and seafood. For example, one 4oz serving of liver provides 3,000 per cent of the UK daily requirement. Plants generally do not produce Vitamin B12: although some organisms (such as certain fungi and algae) produce low levels of it, there are no naturally occurring vegetarian sources that offer a reliable and safe source of B12. Even so, there is no need for *anyone* to be short of B12 in the diet, given the wide extent of fortification in food and the range of B12 supplements available – including Marmite yeast extract. A study at York University found that eating a teaspoon of Marmite a day for a month improved the brain's response to visual patterns by 30 per cent.

But it's not just how much B12 that we eat that matters. The stomach produces a special protein called 'intrinsic factor' which is essential for the absorption of Vitamin B12. Loss of intrinsic factor – for which risk factors are bypass surgery, auto-immune disease such as Crohn's disease and coeliac disease, being an older adult, contracting HIV and a family history of the disease – is the most common cause of Vitamin B12-deficiency anaemia.

Sharp fruit, sharp mind

On 21 October 1805, a naval battle took place that changed the face of Europe for generations – the defeat of the French and Spanish navies at Trafalgar by a British fleet under Admiral Horatio Nelson. Leadership, unparalleled seamanship and masterly gunnery are the ostensible reasons for the victory. But there is also a little-known and seemingly unlikely cause: the superior diet of the Royal Navy. Since 1795 every British sailor had drunk a daily ration of lemon juice – a source of Vitamin C – to prevent scurvy. Their counterparts did not. And we should not underestimate the importance and effects of this deficiency disease. In 1622 Sir Richard Hawkins, another intrepid naval commander, stated: 'In 20 years . . . I dare take upon me to give accompt of 10,000 men consumed with scurvy.'[2] Never before and never since, it has been said, has a dietary supplement had such a crucial role in global politics.

And if a lack of Vitamin C has such a profound effect on overall health, are there any particular consequences for the health of our brains? The answer is yes; and the evidence falls into three categories. First, we now understand the critical role that Vitamin C plays in the chemistry of the brain. It is highly concentrated in our brain cells; it is an anti-oxidant, protecting brain DNA from free radicals; it's essential for the building of the protective myelin sheaths on the fibres that connect brain cells; it converts dopamine into serotonin, and regulates the release of these two vital messenger molecules. Second, there is evidence that it plays a part in our mental capabilities. Between 2010 and 2013, a major longitudinal study of adults in New Zealand (the CHALICE study) found that 62 per cent of 50-year-olds – interestingly, more men than women – had inadequate blood levels of Vitamin C, and that lower levels of mild cognitive impairment were observed in those with the highest concentrations of Vitamin C in their blood plasma. Then, in 2017, Australian researchers published a review of fifty similar studies from the previous thirty-seven years. They found that degree of cognitive impairment was significantly related to blood concentrations of Vitamin C. Among those who were not so impaired, higher blood levels of Vitamin C were associated with higher cognitive ability. Third, many studies have shown that lower Vitamin C intake over the whole of our lives plays a role in the type of neural losses typical of dementia.

The RDA for Vitamin C varies according to age, gender, health status and, rather infuriatingly, the source of the guidance. In the UK, for adults aged 19 and over, the recommended minimum is 40mg a day. In Europe, the European Food Safety Agency recommends 100mg a day for men and 95mg a day for women. It has to be acknowledged that dietary advice is not an exact science – medical opinion varies, as does the certainty of the research on which it is based. Foods rich in Vitamin C are rosehips, berries, bell peppers, kale, broccoli, Brussels sprouts, papaya, citrus fruits, pineapple and cabbage.

So how much should we take? The key is to eat an adequate amount of Vitamin C every day for the whole of our lives. It is the consistency and regularity of intake that matters, once the minimum has been ingested. There are no recommendations specifically for brain health, though some evidence does show a 'dose response' between the level of Vitamin C in the blood and cognitive ability. Exceeding the RDAs shown above may

Fruit on a table

have benefits and certainly is not harmful. There isn't enough evidence to recommend a maximum level (the famous chemist Linus Pauling took 1.8g per day!), and Vitamin C is one of the safest and best-tolerated dietary elements. So munch away freely on your broccoli and Brussels!

Sunbathing your brain

In 2012, in one of the richest societies, and one of the sunniest climates, anywhere on the globe, an unthinkable event occurred. Australia experienced an outbreak of childhood rickets, a disease associated with poverty and caused by a deficiency of the sun-related vitamin, Vitamin D. The consequences for general health are bad enough. Vitamin D is essential for calcium and phosphate metabolism in the body. It is activated by the liver and then travels in the blood to its targets – the small intestine, the kidneys and the skeleton. Vitamin D deficiency results in rickets – failure of the bone to harden – a disease which, hard as it may be to believe, is on the rise in Western economies, an increase almost entirely due to behaviour such as avoiding the sun, covering the skin, prolonged breast-feeding without Vitamin D supplementation and a lack of dairy products in the diet. And it's not just a matter of bone health: if you are short of Vitamin D, there will be consequences for your brain health too.

Recent research has shown that there are receptors for Vitamin D throughout the central nervous system and in the hippocampus. We also know Vitamin D activates and de-activates enzymes in the brain that are involved in neurotransmitter synthesis and nerve growth. Further, animal and laboratory studies suggest that Vitamin D protects neurons and reduces inflammation.

Two new European studies looking at Vitamin D and cognitive function have taken us one step further. The first study, led by neuroscientist David Llewellyn of Cambridge University, divided more than 1,700 men and women, aged 65 or older, into four groups based on their blood levels of Vitamin D: severely deficient, deficient, insufficient (borderline) and optimum. He then tested for cognitive function. Those who had the lowest Vitamin D levels did the worst on a battery of mental tests: the 'severely deficient' group were more than twice as likely to be cognitively impaired as those in the 'optimum' group. The second study, led by scientists at the University of Manchester, looked at Vitamin D levels and cognitive performance in over three thousand men aged 40–79 in eight different European countries. The researchers found that those men with lower Vitamin D levels were slower at information processing, with a particularly marked difference among men aged over 60.

On the basis of research to date, then, we can state with some confidence that low levels of Vitamin D are related to brain performance, to cognitive decline and to the lifelong risk of dementia. But we don't yet know *how* this relationship works; nor do we know yet whether supplementing Vitamin D intake will actually slow a decline that has already started.

Vitamin D is the only one of the thirteen vitamins that we can make in our bodies – in the skin, in response to the action of sunlight. The problem is that most people in the northern hemisphere get very little sun. Most of us work indoors, and most of us live in countries that don't get much sunlight during the winter. The necessary rays are only emitted all year round in latitudes below 35 degrees. It's estimated that about 1 billion people across the world are seriously deficient in Vitamin D, and typical population studies reported in the scientific press suggest that levels are too low in up to 50 per cent of people.

To get enough Vitamin D, then, most people in temperate or northern climes have to find it in their diet, where it comes in a variety of forms. The main two are Vitamin D2 ('ergocalciferol') and Vitamin D3 ('cholecalciferol'). Vitamin D2 is abundant in animal foods, but is found nowhere in the plant kingdom except rarely in 'microalgae' and in small amounts in fungi. Vitamin D3 is found only in animal foods, notably fatty fish, fish oil, eggs, butter and liver. The UK and US government RDA for Vitamin D is 15mcg a day; the European Food Safety Authority sets a guideline of 600IUs. Some experts, however, set the target much higher, at between 1,000 and 2,000IUs. To generate even the lower of those two figures, 1,000IUs, would take 15–30 minutes of exposure to the sun two to three times a week – not a practical option for most of us living above 35 degrees north!

Wherever we get it from, what matters for our health is the amount of Vitamin D circulating in the blood – known as 'calcifediol' or, in strict chemical terms, 25(OH)D. Wherever possible, we should be eating Vitamin D3, because research has shown that dietary Vitamin D2 is less effective than D3 in raising blood levels of calcifediol. For example, one study of thirty-two women aged between 66 and 97 found that a single dose of Vitamin D3 was nearly twice as effective as Vitamin D2 in raising calcifediol levels. In other words, if you consume only Vitamin D2, you are likely to have only half as much calcifediol circulating as if you consume only Vitamin D3.

It is difficult in any case to consume adequate amounts of Vitamin D2 from non-animal sources. For example, the Vitamin D2 content of the kind of fresh button mushrooms you'd buy in the supermarket is commonly reported to be less than 1mcg per 100g (4oz) fresh weight, so you'd get only a negligible amount from that typical 100g serving – especially given that cooking destroys about 40 per cent of the bio-available Vitamin D2 in mushrooms. So those who choose not to eat food sourced from animals may need to take a Vitamin D supplement to protect brain health.

Beautiful skin, beautiful brain

If sunlight increases the formation of Vitamin D in the skin, there is a price to pay: exposure to the sun reduces its Vitamin E content. As does age. And if you are interested in having beautiful skin, Vitamin E is your best friend. The exposed skin of the body is subject to constant ultraviolet irradiation and oxidative damage. Vitamin E is the principal, naturally occurring, fat-soluble anti-oxidant that protects skin from the adverse effects of oxidative stress, including the ageing effects of light. But it is not just the skin that suffers from these effects.

Free radicals are common in the brain – the price we pay for its high oxygen consumption. Nowhere in the body is the need for anti-oxidants as acute. It is therefore not surprising that Vitamin E, a powerful anti-oxidant, is increasingly shown by the evidence to be important to brain health. One of the earliest studies, carried out in Chicago, published its results in 2002. Involving more than four thousand adults over a period of three years, it showed that a higher intake of total Vitamin E was associated with less decline in cognitive score per year. It appears that over a lifetime a diet with a low intake of Vitamin E ages our brains by eight to nine years compared to one with a high intake. Or, to put it another way, if we consume a lot of Vitamin E we can effectively delay brain ageing (in terms of cognitive decline) by eight to nine years. These findings were confirmed by another study carried out ten years later at the Karolinska Institute in Sweden. This one looked at 187 adults, some with mild cognitive impairment and some with Alzheimer's disease. It found that low blood plasma levels of all forms of Vitamin E (there are eight) were associated with an increased likelihood of both mild cognitive impairment and Alzheimer's.

And how does Vitamin E protect the brain? It's intriguing. The lipid (fatty) layers of tissue that surround our brain cells are subjected to 'oxidation fires' thousands of times a day; and Vitamin E is the 'firefighter' that puts them out – a task it accomplishes with ease because it is fat-soluble and so it can dissolve itself in the fatty outside layers of the cell, therefore entering it easily.

The RDA for Vitamin E (all forms) is agreed by all authorities to lie between 7 and 15mg. Estimates from a global review of over a hundred studies showed that the average daily intake was just 6.2mg – well below the recommended level and a potential brain health disaster in the making. Food sources rich in Vitamin E include wheatgerm oil or wholewheat grains; plant oils (such as olive oil, rapeseed oil, sunflower oil etc.); many nuts (e.g. almonds, peanuts – including peanut butter – cashews and hazelnuts); avocados, spinach, asparagus and broccoli. Animal sources include fish, oysters, butter, cheese and eggs.

Increasing evidence shows that Vitamin E taken in supplement form is less effective in providing health benefits than natural Vitamin E ingested in wholefoods – and sales of Vitamin E supplements are declining as a result. Supplements contain only one of the eight naturally occurring molecules in the vitamin – and even this one, alpha-tocopherol (E307), is twice as potent in natural foods as in supplemental form.

Eat your greens

It's only in recent years that researchers have investigated the influence of Vitamin K (actually a group of vitamins) on brain health. These studies have discovered that Vitamin K plays a vital role in brain chemistry, is found in high concentrations in the brain, and has a relationship with both cognitive performance and brain health. First, the chemistry. Vitamin K is anti-inflammatory and anti-oxidant, and is involved in sphingolipid chemistry. (The gloriously named sphingolipids are fats that help the growth and survival of brain cells.) Second, brain performance. Changes in this have been shown to occur in small mammals deprived of Vitamin K, but only when the deficiency is prolonged. In one study, rats showed learning problems in a maze – but only after some twenty months of Vitamin K deficiency, which is nearly halfway through the four-year life of the average lab rat. The rats that were deficient in Vitamin K at six and twelve months of age did not show any learning problems. What about humans? In 2019, a team of Italian scientists took a careful look at eleven studies that investigated the adult human brain and Vitamin K. In seven of them – all of which examined participants over age 60 – they found an association between Vitamin K deficiency, mental performance (e.g. verbal memory) and cognitive decline. For those where participants were aged under 60, there was no apparent relationship. An earlier review paper produced evidence that Vitamin K slows down the onset of cognitive impairment. It must be emphasized that these studies do not show cause and effect, only an association. Nevertheless, the evidence suggests that an adequate lifelong dietary intake of Vitamin K is important to brain health.

Of the two main forms of Vitamin K, one is found naturally occurring in leafy green vegetables, such as spinach; the other is manufactured in the colon by the gut bacteria (the microbiota). The physiological contribution of colon-manufactured Vitamin K remains uncertain, and because of this and other unknowns, dietary advice is given in terms of 'adequate intake' rather than as an RDA. Most authorities recommend 1mcg per kilogram of body weight as a daily intake – which isn't very helpful. As a general rule, we can say that adult women should be eating 55mcg a day and adult men 65mcg a day. The main sources in a Western diet are green leafy vegetables (spinach, kale, broccoli, lettuce) and soybeans; box 5.4 shows what you can expect to get from a typical serving of each.

BOX 5.4: FOOD SOURCES OF VITAMIN K

Food	Mcg in 100g serving	% of daily recommended intake
Spinach, raw, 1 cup (240mls)	145	121
Kale, raw, 1 cup (240mls)	113	94
Broccoli, boiled, ½ cup (120mls)	110	92
Soybeans, roasted, ½ cup (120mls)	43	3
Iceberg lettuce, 1 cup (240mls)	14	12

Source: US Department of Agriculture Food Composition Database.

Magnesium for memory

In 1905, only 1 per cent of Americans had depression prior to the age of 75. In 1955, this had increased to 6 per cent of Americans *before the age of 24*. In the fifty intervening years, there was a sea-change in food production methods. The consumption of wholegrain foods plummeted as the food industry refined grain to produce white flour and white bread, thereby slashing the mineral content of baked products. For example, in 1905, people were getting about 400mg of dietary magnesium from bread every day; but by 1955, white bread was the norm – and it was almost completely devoid of magnesium. Magnesium might seem an unlikely suspect for increased levels of societal depression; but a look at its role in the brain will show us why it matters.

Called by some the original 'chill pill', magnesium has been a home remedy for stress and anxiety for decades – indeed, many Victorian spas offered magnesium-rich water for therapeutic bathing, and even drinking, as prescribed by doctors at the time. Modern science now tells us that magnesium is critical to many chemical pathways in the brain, playing a vital role in the production of important transmitter substances, such as serotonin, and balancing the excitatory effects of calcium on the brain. And there is a twist: magnesium is vital for the conversion into active form of several B vitamins – vitamins that are essential for brain health. Many vitamins need to be modified in this way to fulfil their function in the body's chemistry, for many different reasons: for example, the active form may not be present in the food, or slight chemical changes may be needed to enable absorption into the cells. Animal studies have shown that magnesium helps us to learn and remember – and protects our memory over time. It has even been asserted that magnesium intake (measured by its presence in hair) is related to academic performance. Research has also shown that magnesium suppresses the release of the stress hormones, such as cortisol, and moreover acts on the blood–brain barrier (a membrane in the blood vessels which prevents the entry into

The Bean Eater

the brain of many circulating products) to deny cortisol access to the brain. The paradox is that during stress magnesium is washed out of the body. Our ancestors would have had a ready source of resupply in a diet rich in natural meat and seafood. Today, our food generally contains little magnesium, and there isn't more than a trace in either bottled or tap water. The current RDA for adults (women and men) is 320–420mg in the USA and 270–300mg in the UK, but we struggle to take in much more than 250mg daily. Magnesium-rich foods (soil dependent: that is, depending on its being present in the soil where they are grown) include almonds, spinach, cashew nuts, peanuts, soya milk, black beans, kidney beans, bananas, raisins and brown rice.

Keeping the conversation going

I hesitate to use the term 'super nutrient', but zinc is a clear candidate. It is a powerful anti-oxidant, protecting DNA against attack by free radicals. It sustains the life of cells, including those in the brain. And it slows down ageing through its anti-inflammatory properties. Incidentally, for the men among my readers, it is also essential for maintaining an erection. More than fifty years ago, scientists discovered that the brain contains high concentrations of zinc. We now know that it is particularly concentrated in the vesicles of nerve cells, little 'bladders' that package up those crucial neurotransmitters – the molecules that enable neurons to talk to each other. Now the really interesting point. The highest concentrations of brain zinc were found among the neurons of the hippocampus, the centre of learning and memory. In a remarkable breakthrough, scientists at MIT and Duke University were able to watch zinc in action. They found that binding the zinc to proteins and reducing mobile 'free' zinc prevented communication between two critical groups of cells in the hippocampus. Without free zinc, the cells *stop talking to each other*.

Given what we know about zinc, then, it should not come as a surprise to find that animal studies have shown zinc deficiency to affect neurogenesis and increase the rate of cell death in the brain, which can lead to learning and memory deficits. But there is currently a dilemma. Some studies report a decrease in zinc in the Alzheimer's brain – but others have found an apparent connection between Alzheimer's and aberrant high zinc levels promoting plaque formation. Only more research will resolve this paradox, which serves if nothing else as a subtle warning about supplementation. We can't take it for granted that more is better. Moderation is the key.

Zinc inadequacy is common both in Europe and the USA, with about 20 per cent of adults not getting enough from their diets. This has raised concerns about public health. The body cannot store zinc, so we need to eat it every day: about 9mg for women and 11mg for men. Foods especially rich in zinc are oysters, crab, lobster, beef and chicken. If we regularly eat these foods then deficiency should not be a problem. Zinc is found in only very small amounts in a vegetarian diet, in foods such as beans, chickpeas, nuts and pumpkin seeds (all soil dependent). Supplementation, however, is not advised unless you have been shown to be zinc deficient by a doctor.

A fire in our head

Our adult brain weighs about 3lb, and nearly two-thirds of it – about

60 per cent – is made of fat, once the water content is set aside. Of this, omega-6 fats make up about 2 per cent and omega-3 fats make up over 80 per cent. Both of these are unsaturated ('good') fats. The remainder is saturated fat. We have evolved so that our bodies can make saturated fats but not unsaturated ones. So we have to eat the latter every day.

But there is a twist in this story. Over the course of human history the fat content of our diet has increased enormously, and the type of fat we consume has changed too. In our hunter-gatherer food, the ratio of omega-6 to omega-3 fats was about 1:1 – that is, we ate equal amounts of both. The ratio is now 20:1 – in other words, we eat twenty times as much omega-6 fat as omega-3 fat. This is largely because we eat many more processed foods which contain omega-6 fat from vegetable oil. Our intake of omega-6 has risen from 1–2 per cent of daily calories before the 1930s to 7 per cent today.

And now the key point: most of this omega-6 that we eat *is not integrated into the brain*. It's oxidized into caustic by-products such as carbon dioxide, acetate and other highly inflammatory substances. *We are lighting a fire in the brain*. Research has shown that if we reduce the omega-6 ratio we may well protect the brain against inflammation. Scientists have even found the magic ratio: we can maximize the anti-inflammatory effect by keeping the omega-3 to omega-6 ratio at 4:1.

Enough of omega-6. What of omega-3? Research has shown that the two particular omega-3 fatty acids we need to have floating around in our brains are called DHA and EPA. DHA alone makes up 40 per cent of all the fatty acids in our brain, where it is highly chemically active and absolutely essential for brain development and maintenance. Its special structure enables efficient signalling between neurons, facilitating the complex brain processes responsible for intelligence. Having adequate levels of DHA and EPA protects brain function and brain health. Both are indispensable to our body's chemistry, moderating blood clotting and combating inflammation and its effects, such as swelling and pain. Research has found that increased circulating levels of EPA and

DHA are associated with a lower risk and severity of cognitive decline, dementia, depression and brain atrophy. Overall, research favours diets rich in these two 'long-chain' omega-3 fatty acids. It's also worth mentioning one more, called DPA – for three reasons: we need it to store EPA and DHA; it's hugely important for our nerve cells; and its blood levels are extremely low in most people. You can find it in fish oil and grass-fed beef.

We eat plenty of omega-6 fatty acids, easily enough to make up the recommended 7 per cent of our daily energy needs. But we do not eat enough omega-3 oils, especially DHA and EPA. The 'adequate intake' of omega-3 for adults is 1.1 grams per day for women and 1.6 grams per day for men. So what should we be eating to make sure we get this?

We've already seen that two things in this area are essential for brain health: high levels of circulating DHA and EPA, and a low omega-6 to omega-3 ratio. And therein lies the problem (or rather, three problems in one – especially for vegetarians). Plant foods have low DHA and EPA content; the high omega-3 they do have (called alpha-linoleic acid or ALA) converts poorly into DHA and EPA in our bodies; and the typical omega-6 to omega-3 ratio in plant foods is awful. In almonds, for example, it's an astonishing 2,000:1, and in cashews it's 200:1. If we look at animal foods, a 4oz (100g) serving of oily fish (herring, sardines, salmon, trout, tuna or crab) will give us the daily adequate intake of DHA and EPA and an exceptionally good ratio of omega-6 to omega-3. In tuna, it is 1:30 – completely reversing the 'bad' ratio we saw earlier, and more! So the questions we should ask are: how many of us get a serving of oily fish (or fish-oil) every day? And how many of us rely on plant sources for the protective DHA and EPA? Now for the bad news on supplements. An expert review conducted in 2016 by the respected Cochrane organization, whose purpose is to facilitate evidence-based decisions in health, showed no beneficial effect of omega-3 supplementation on cognitive function or dementia severity. We are talking here about synthetic, industrially produced supplements with no natural history. Research has shown that synthetic supplements are absorbed differently from, for example, vitamins in fresh food. How the body uses them is also suspected to be different. One reason is that fresh nutrients are eaten in diverse chemical combinations, which is not true of isolated supplements. So: eat fresh!

Send for chocolate

No chapter on nutrition and brain health would be complete without a mention of flavonoids – naturally occurring plant pigments that are found widely in fruits and vegetables. In animal studies they have been found to have multiple benefits for the brain, and particularly for cognition. They are anti-oxidant, suppress inflammation, protect neurons against toxins, promote neurogenesis, and enhance learning and memory. Furthermore, we have a fair understanding of how they operate within the brain. They protect neurons against injury by toxins, they suppress neuro-inflammation, they improve blood flow and they regulate how genes are expressed.

We also have some research which shows that flavonoids have benefits for mood and memory in humans. However, few clinical intervention studies have been done, and there are currently no data in support of a *causal* relationship between eating flavonoids and behavioural outcomes in humans. They are poorly absorbed in the gut and are quickly used up in the small and large intestines – but this is probably their big secret. They feed the bacteria in the gut, which as we have seen are vital for brain health. The effect of flavonoids depends on prolonged consumption, so we should eat them regularly throughout our lives. Foods high in flavonoids are parsley, onions, all berries, black and green tea, bananas, all citrus fruits, ginkgo, red wine and cocoa. Send for chocolate!

Drink before you're thirsty

You are in the office. It is hot and the air is dry. The sun streams through the window. You feel tired, fatigued even, and your head aches. You can't concentrate and you are slightly dizzy. You maybe had a few drinks the night before. Chances are *you are dehydrated*. Aspirin won't solve it – and neither will tea or coffee, because they are both diuretics, so will just make you more dehydrated. The brain is exceptionally sensitive to water losses. It is in fact about 80 per cent water, whereas other tissues in the body are generally about 60 per cent water. So even small water deficits will impair your mental sharpness.

Dehydration is commonly defined as a loss of more than 10 per cent of total body weight through water loss. We know that this level of water deficiency is bad for brain function and, if it is persistent, bad for brain health too. Most studies have found that at all ages, dehydration has negative effects on both mental performance, such as attention and short-term memory, and mood. And the deficits arrive quickly: within two hours of the onset of dehydration. In one recent small study, a group of male students aged 19–24 were dehydrated for thirty-six hours and then underwent a series of tests on aspects of mental performance, including mood, attention, memory and reaction speed, for thirty minutes. At the end of this period, they were allowed to rehydrate, and were then retested. While the participants were dehydrated, their mood scores were lower, their fatigue scores were higher, and their attention, memory and reaction speed were all significantly impaired. After one hour of rehydration, performance in all measures was restored. Interestingly, studies show that women of all ages tend to be more markedly affected by dehydration than men. One explanation is the lower percentage of muscle in women, which may make them more sensitive to dehydration (fat tissue is only 10 per cent water, and women on average have 20 per cent more body fat than men). While we know that dehydrated neurons are associated with brain dysfunction, it is not yet known whether dehydration is a cause or an effect of age-related cognitive decline. As we get older, we are less able to regulate salt and water in the body, and less sensitive to thirst and the need for water generally – and so tend to drink less.

Thinking practically, then, how do we know when we are short of water and how much should we drink? First, do not rely upon thirst – it is a very poor indicator of water levels in the body, arriving late and leaving early. As a guide, by the time we have a compelling thirst, we already have a water deficit of 1–1.5 litres (2–3 pints). A better guide is the colour and appearance of urine. In hot conditions, the US Army encapsulates its advice to soldiers in a brilliant expression: 'Pee white once a day.' The darker the urine, the lower the water content of the body. It's vital to know and obey the rules on fluid intake. As a general rule, to keep the brain properly hydrated women should drink 2–3 litres (3.5–5 pints) of fluids every day, and men 2.5–3.7 litres (4–6.5 pints). There is also some evidence to support the anti-oxidant and anti-inflammatory health benefits of 'alkaline ionized water', with its high levels of dissolved hydrogen and 'microclustered' water molecules. Ionizing water filters are easily available to buy, and millions of people across the world use them in the home to ionize tap water. The water simply passes through an activated carbon filter to purify it and then between two platinum electrodes to 'electrify' or ionize it.

Challenge your brain

Chris Talley – whom we met earlier in this chapter as the organizer of Luke Aikins' amazing B6-zeroing skydive – is an expert nutritionist of some repute. He coaches many of America's leading athletes, members of the military special forces and elite performers – all individuals working in areas where success or failure often depends upon 'marginal gains', the tiny edge that can make all the difference between winning and losing. Under great stress. With the 2021 Olympics on the horizon, I asked Chris about the difference that good or bad nutrition can make to our 'mental edge' – keeping our concentration and recall, and maintaining emotional control while dealing with both acute and chronic stress.

I started by asking Chris whether, had he known that Tom Brady, the National Football League's leading quarterback, would be performing at that level for twenty-five years, he would have advised him any differently. He responded:

Because Tom Brady has paid attention to nutrition, he hasn't been eating a lot of junk. More than his ideal diet, it would be the lack of junk coming in that has contributed to his longevity from a nutritional perspective . . . You can close the mental edge by doing things with nutrition. My aim is to make a 1 per cent difference which doesn't sound very much but then people will say they feel 100 per cent better . . . and the difference between [Olympic] gold and silver is less than 1 per cent.

I then asked Chris about 'staying sharp', which is the number one concern for many people as they get to their forties. He said:

I would say omega-3s play a significant role in keeping sharp. Get in grass-fed beef, fatty cold-water fish or even cod liver oil. I know it sounds old school and it's the worst thing I've ever tasted but it gives you the omega 3s DHA and EPA (the molecularly distilled cod liver oil doesn't taste bad). I've heard vegans say they don't have to take fish oil and can get their omegas from alpha-linoleic acid in nuts and seeds. That's true to an extent but when there's any sort of a demand, the conversion rate to produce DHA, DPA and EPA from ALA is too slow. I've seen the blood panels and they don't end up with these other omega-3s . . . I've seen over five thousand [blood panels] and if you asked me for the worst ten, nine of them would be vegan . . .

I am a vegetarian for ethical reasons and I am probably the only vegetarian saying I would be healthier if I ate meat – but it's true. I get to see my own blood panels so I know what to do, to work around that . . . but most people do not have this [advantage].

My next question to Chris was: is there any general advice he would give to ordinary people, often with stressful and demanding jobs, who wish to learn from athletes? He replied:

The greater the stress, the more people rip through certain micronutrients, like B6 and magnesium. In 'Heaven Sent', he [Luke Aikins, the skydiver we encountered earlier in this chapter] ended up just wiped out, anything associated with B6 was gone, so it pays to build up B6 beforehand if you know a crisis is coming . . . If you are looking to put a stress meal together, try fresh red bell peppers, sweet potato, grass-fed beef with pistachio nuts and sunflower seeds.

You need both B6 and tryptophan, an essential amino acid, to make serotonin, so if you are low on these there is a big negative impact, not just on sleep but on mood as well . . .

If a stress biomarker like CRP [C-Reactive Protein] is high, that will have a negative impact on long-term brain health. Anything to reduce it [CRP] is going to be a huge advantage. From a nutritional perspective, reducing inflammation would drop it, so something as simple as curcumin would do it.

Finally, I asked Chris about taking supplements. His response indicates caution:

I aim to advise not to have supplements but there are situations where if some people just aren't going to eat certain food groups, then you really don't have a choice. So you do end up with cases where supplements apply . . . but I think a lot of these people who are taking a ton of supplements have no idea what they're doing and in fact they may be causing more damage than good . . . In one case where we were taking stool samples, one of the things we saw was a pressed pill supplement tablet just the same as when they took it!

Working out a nutritional plan

This chapter is *not* about 'diets'. Dieting and diets are a hugely complicated issue. There are hundreds of diets available. Everyone's personal needs are different, and individual psychology plays a huge part in sticking to them. And good-quality evidence on what works and what doesn't is hard to come by.

What we can say is that combining a good variety of types of food and nutrients in our diets is the route most likely to deliver health benefits for our brain. There is no single food that acts as a silver bullet for improving or maintaining brain function. And it is our lifelong eating habits that promote brain health. We can benefit from changing to a healthy diet at any age; however, the sooner we start, the better. Long-term healthy eating habits promote good brain health.

Having said all that, there are three particular diets which I would like to bring to your notice because they have all accumulated a certain amount of research evidence in their favour. They are the Mediterranean diet, the DASH diet and the MIND diet.

Let's take the Mediterranean diet first. Obviously, what people eat in this region varies among and within individual countries, but in all of them the diet is high in vegetables (including legumes), fruits, nuts, beans, cereals, grains, fish and unsaturated fats such as olive oil. Intakes of meat and dairy food are usually low. The Mediterranean diet has been shown to improve cognitive function, with moderate evidence for a reduced risk of cognitive decline with age. These studies include some carried out in the Blue Zones discussed in chapter 2, where everyone eats either a Mediterranean diet or something very close to it. In 2019, a rigorous review paper by Annelien van den Brink and colleagues in the Netherlands found that closer adherence to the Mediterranean diet was associated with better cognitive scores in nine out of twelve cross-sectional studies, seventeen out of twenty-five longitudinal studies, and one in three trials. A Mediterranean-type diet is also recommended by the Global Council on Brain Health. Their recommendations are summarized in box 5.5.

BOX 5.5: DIETARY RECOMMENDATIONS OF THE GLOBAL COUNCIL ON BRAIN HEALTH

For individuals

A. Encourage:	B. Include:	C: Limit
• berries (not juice) • fresh vegetables (in particular leafy greens) • healthy fats (such as those found in oils, including extra virgin olive oil) • nuts (a high-calorie food, so limit to a moderate amount) • fish and seafood	• beans and other legumes • fruits (in addition to berries, previously mentioned) • low-fat dairy, such as yogurt • poultry • grains	• fried food • pastries • processed foods • red meat • red meat products • salt

1. We recommend the food guidelines in the table above for brain health. We encourage people to eat the 'A-list' healthy foods regularly, include other 'B-list' foods in their diet, but to limit the amount of 'C-list' foods.

2. If you don't drink alcohol, don't start drinking in order to protect your brain health. If you drink alcohol, do so in moderation, because it is unclear whether there is any beneficial level of consumption for brain health.

3. Eat whole, non-processed foods to limit unintentional intake of too much salt, sugar and saturated fats, which often appear in processed, packaged and fried foods.

4. Be cautions when it comes to eating chocolate. Cocoa-rich products are generally high-calorie because they often include sugar and high-fat dairy products. Therefore, when incorporating chocolate into your diet, it is important to avoid excess weight gain, which could counterbalance, or even exceed, any benefits from eating cocoa.

5. Avoid trans fats.

Now the DASH diet. 'DASH' stands for Dietary Approaches to Stop Hypertension. The diet is really a lifelong approach to healthy eating, and is designed to help treat or prevent high blood pressure (hypertension) by emphasizing a low intake of sodium and high intake of nutrients that help lower blood pressure, such as potassium, calcium and magnesium. But there is reasonable evidence that long-term adherence to a DASH-type diet is also beneficial to brain health. The van den Brink paper I just mentioned found that adherence to the DASH diet was associated with better cognitive function in one cross-sectional study, two out of five longitudinal studies and one trial.

As for the MIND diet, this acronym stands for Mediterranean–Dash Intervention for Neurodegenerative Delay. This approach draws on both the diets just mentioned above, and was specifically constructed on the basis of research evidence on maintaining brain health. The Dutch review paper found that higher adherence to the MIND diet was associated with better cognitive scores in one cross-sectional study and two out of three

A well-furnished table

longitudinal studies. One of the studies was conducted by Kristine Yaffe's team at the Brain Health Institute in California. They found that in their sample of nearly six thousand older adults (average age 68), those who stuck to a Mediterranean diet or a MIND diet had better brain function and lower risk of cognitive impairment. But the MIND diet is *not* a diet for 'older' people: its preventative value begins in early middle age. Nor is it a highly prescriptive diet: it simply suggests that we eat more of ten recommended foods and less of five that should be limited. These categories are the same as those shown in box 5.5, with the exception that under 'C: Limit' the GCBH has included processed food and salt as well as the five foods specified by the MIND programme.

Should you include supplements in your nutritional plan? Brain-health supplements generated $3 billion in sales globally in 2016, a total projected to reach $5.8 billion by 2023. There are many reasons for this public obsession. One of them is certainly the unrelenting and often misleading advertising of claimed benefits from supplements by an industry which exploits people's fears about health. In fairness, though, we should also note that dietary intake of micronutrients does not necessarily translate into adequate levels in the blood; and further, that some nutrients, such as Vitamin D, are only found in a narrow range of foods. In these cases, and for good *general* health, government agencies do recommend supplements.

So what does the science say about supplements specifically for brain health? In 2018, the UK Cochrane Database of Systematic Reviews published a global study of all available evidence on supplements and the brain, involving studies on the B vitamins; on the anti-oxidant vitamins A, C and E; on D3 and calcium; and on selenium, zinc and copper. This body of research involved a combined total of 83,000 participants. The aim of the review was to find out whether people aged forty or over could maintain their mental abilities or reduce their risk of cognitive decline by taking vitamin or mineral supplements. The conclusion? I quote: 'We did not find evidence that any vitamin or mineral supplementation strategy for cognitively healthy adults in mid or late life has a meaningful effect on cognitive decline or dementia.'[3] There is a caveat: the research available is incomplete. However, the key messages remain:

- For most of us, the best way to get nutrients for our brain health is from a healthy diet and not via supplements.

- Because dietary supplements are subject only to general food law and are not assessed on the truthfulness of their claims, we should view claims made on supplement packaging and in marketing materials with scepticism.

- Supplementation may be beneficial for people with lower than recommended levels of micronutrients such as Vitamin B12 or Vitamin D; however, remember that the only valid means of detecting deficiencies in the absence of clear symptoms is via a blood analysis, which you may request from your doctor.

And it's worth remembering the statement from the Global Council on Brain Health: 'We do not endorse any ingredient, product or supplement formulation specifically for brain health, unless your health care provider has identified that you have a specific nutrient deficiency.'[4]

We'll end this chapter on some practical and strongly evidence-based advice to enable you to change to a brain-friendly nutritional plan.

- If you're already eating a healthy diet like the Mediterranean, DASH or MIND diets: great! Keep it up. If not, embark on a new journey – move away from the path you are on and towards one of these patterns, so that you're eating mostly plant-based foods plus a little fish and red meat from sheep, goats or grass-fed cattle. In choosing your fish, crustaceans and molluscs, go for those with a high omega-3 and Vitamin B12 content (e.g. salmon, anchovies, sardines, cod, sea bream, trout and prawns). Don't eat too much protein (about 0.7g per kilo of body weight per day is about right), except if you are over 65, when you should increase your protein intake from e.g. fish, eggs and white meat. Even then, however, keep most of your protein coming from plant sources, such as beans, chickpeas and green peas.

- Reduce your intake of saturated fats from animal and vegetable sources (red meat, cheese) and avoid refined sugar. Increase your intake of 'good' fats (omega-3) and complex carbohydrates with high fibre content. For example, eat wholegrains and lots of vegetables (tomatoes, broccoli, carrots, legumes etc.) with generous amounts of virgin olive oil and small helpings of raw nuts (about 25g per day).

- Focus on small steps. Changing lifelong eating habits can seem an overwhelming task, but even tiny shifts can have an effect. For example, increasing fruit intake by just one serving per day has the estimated potential to reduce cardiovascular mortality risk by 8 per cent – the equivalent of sixty thousand fewer deaths in the United States, fourteen thousand in the UK and 1.6 million globally, every year. The brain is no less sensitive to such small but beneficial changes. And what is good for the heart is good for the brain.

- The body thrives on regularity, notably adherence to a consistent daily routine across the 24-hour cycle (much more on this in chapter 9 on sleep). Try to respect this and to eat regularly at the same times, if possible. Bear in mind that the gut has its own rhythm, with times when it should be active and times when it prefers to be resting – and this rhythm is synchronized not only with those of other digestive organs like the pancreas and the liver but also with the hypothalamus in the brain. Try to confine all eating within a twelve-hour period – for example, between 7 a.m. and 7 p.m. And try not to eat anything within the three hours before you go to bed.

- Hydrate constantly throughout the day. Drink more water, and avoid drinking excess tea, coffee and carbonated drinks.

- Stay physically active to complement your healthy diet. As we saw in chapter 2, physical activity has been shown to be beneficial to brain health in adults, and is important alongside a healthy diet in promoting healthy ageing. Your diet should provide the food, nutrients and energy you need so that you can maintain a good balance between food and physical activity.

- Avoid 'junk food' – too much fried food, too many pizzas and too many take-away or precooked meals, especially those consisting of overprocessed or highly refined ingredients. For example, try eating at least one meal a week with fish that is not fried. Buy fresh food with a high mineral and vitamin content and prepare and cook it at home, if at all possible. This gives you control over the salt, sugar and fat content of what you are eating.

- Check out whether the fresh produce that you are eating is really fresh. Much farm produce on sale in shops has spent many days in transit and/or storage. Remember that many micronutrients in harvested food, especially vitamins, decay very quickly. Contrary to common belief, the nutrient value of canned or frozen food is often much higher than that of much so-called fresh food. So select a variety of dietary ingredients – just as our ancestors did, though of course the choice is rather different these days!

- Remember: the best way to feed the brain is to eat a balanced and healthy diet and not to rely on supplements. Spend your money wisely.

The final word

Our modern Western lifestyle of constant work and persistently high stress has taken us away from the feeding patterns of our ancient ancestors, consigning us to prolonged, pressurized mental activity, limited physical exercise, constant eating and snacking – and the consumption of excess calories. But while our habits have changed, our physiology – as we saw in chapter 3 – is largely still geared to a 'feast and famine' energy intake characteristic of our pre-Neolithic ancestors, who endured periods of fasting and had a diet hugely more varied than ours. This clash between our modern behaviour and our ancient physiology is a recurring problem for us if we wish to eat to look after our brains. If we heed our current knowledge of nutrition and brain health – on calories and nutrients – we will not need to wait for our physiological evolution to catch up with our lifestyle. It is the core purpose of this book to encourage you to heed that knowledge – and to do so now.

No brain is an island

IN THIS CHAPTER, WE'LL look at the extent to which our social nature as humans underpins our general health – and especially the health of our brain – as revealed by scientific investigation. And we'll start in an unusual setting: outer space.

Life on Mars

In 2010 in Moscow, in the intriguingly named Institute of Biomedical Problems, an ambitious experiment began which was to change for ever our perception of the risks of space flight. Most importantly, it emphasized the uniqueness – and fragility – of one of the fundamental properties of the brain: its dependence for normal functioning on *meaningful contact with others*.

Mars 500 was a four-year psychosocial investigation of how to prepare astronauts for the long journey to Mars and back – and the hazards they would face on the way: three different gravitational regimes (Earth, outer space, Mars), cosmic radiation, resupply challenges, poor communications and a precarious environment inside their spaceship. As it turned out, the greatest hazard the astronauts would face was none of these. There was one other critical factor which was grossly underestimated, only becoming apparent at the end of the five hundred days of simulated space travel, during which the astronauts did not associate with one another – the effect of social isolation within an already isolated community.

On the simulated outward journey, all seemed to go well with the six-man international crew. But on the return journey, problems arose. By the end, out of the six highly trained, highly intelligent astronauts, each of whom had gone through a rigorous selection procedure, only two had maintained a normal pattern of behaviour. The other four had developed emotional and mood problems – lassitude, social withdrawal ('hibernation') and insomnia – to the extent that they became 'mission incapable'.

Humans and chimps

It would be easy to believe that our higher brain functions, such as learning, memory and reasoning, are divorced from the conditions of our social lives. A look at our evolutionary ancestry, however, will tell us otherwise. While studying the infants of humans and young primates (chimps and orangutans), a group of US and German anthropologists made an unexpected discovery. When it came to exploring the physical world, chimps and humans had very similar cognitive skills. But when it came to dealing with the social world, the human children had much more sophisticated cognitive skills than those of either of the other ape species. The difference was particularly marked in one particular ability, which also varies widely among individuals – the capacity to read what is going on in other people's minds, or what we call 'intuition'. This peculiar ability is not some trivial social convenience. The human brain is hard-wired to support one overridingly important purpose – the survival of the species. For a social species like humans, life on the social perimeter was not only unpleasant and unsafe – it was a threat to survival. And so, in evolutionary terms, the pressure of social isolation and the excruciating pain of loneliness drove us to maintain the connections we needed to ensure survival and to promote social trust, cohesiveness and collective action. The group is EVERYTHING. Without social cohesion – working as a group – humans would not have survived the harsh reality of nature, particularly the imperative of getting enough energy through their food supply. The life of the hunter-gatherer is a precarious existence. Not only food but successful reproduction hinges upon the group. Human children are dependent for longer after birth than any other primates. Human young need protection to survive to maturity. It is the group that provides it; and our brain has evolved that special facility – social cognition – to enable humans to work successfully together.

Our capacity to think, reason, learn, plan and predict is rooted in the social evolution of our brains. It should therefore not surprise us to learn that science is discovering that damage to our social lives influences our ability to think and reason – and even affects the structure of the brain.

Even for members of animal species which we could not expect to have any perception of loneliness, social isolation has adverse consequences. From *Drosophila* fruit flies to mice and rats to agricultural animals, individuals do not thrive when isolated from the group. Cacioppo's milestone paper of 2014, 'Evolutionary mechanisms for loneliness', provides us with a litany of examples: mice that become obese and diabetic; rats that derive no benefit from running; rabbits that show increased levels of stress hormones; and squirrel monkeys that lose their morning rise in the stress hormone, cortisol. All from being isolated. (We should not, by the way, underestimate the importance of early morning cortisol. We may think of it as a 'stress hormone', but cortisol is vital to wake us efficiently and prepare us for the rigours of our day.)

So, the question must be: as humans, do we see such harmful effects in ourselves? The answer is yes; much the same physiological events occur in our bodies if we become isolated. But there's one tangible and subtle difference. We collaborate and cooperate in ways that other primates do not. For humans, what matters is not just whether we are surrounded by others; it's our perceptions, our evaluation of the other humans around us. Humans can be loyal, friendly, compassionate and empathetic. They can also be murderous, hateful, treacherous and deceitful. In the words of Michael Tomasello of the Max Planck Institute in Germany, we are 'ultra-social', with an enhanced capacity for social cooperation. Our sophisticated ways of collaboration changed the way that early humans procured food – but they also changed the way we understood ourselves in relation to others. It is the *quality* or meaningfulness of our company, rather than the *quantity* of our interactions, that is of critical importance to us. All the research shows this. And there's more: over and above what we know about the potential damage resulting from 'objective' isolation, there is the additional damage resulting from *perceived* isolation or loneliness. So we must draw a hugely significant distinction between, on the one hand, social (or 'objective') isolation, where a person has little or no contact with others; and, on the other, loneliness, which is an emotional state, an often excruciating subjective and anxious feeling about a lack of connection with others.

In summary, we are creatures of our social evolution. We are driven by psychological fear of isolation and loneliness, as much as by our innate biological needs, not only to be members of a group but to have

Loneliness

meaningful social relationships within it. We need social bonds as much as we need food and water. Our sophisticated brains have been honed by millions of years of natural selection to work exquisitely as members of a purposeful, rapacious, unrelenting group, making us the deadliest, meanest and most feared species on the planet – and, paradoxically, also the most compassionate. Our psychology, behaviour and social structures have all evolved together, generating a formidable combination of survival assets. Predatory, cunning and grimly articulate, we can out-manoeuvre our enemies or prey, human or otherwise, *by working as a group*. Aristotle did not need science to pronounce us social animals. But science is now revealing the extent to which our social nature underpins our health – including the health of our brain.

The myth of the loneliness epidemic

Before we look at how our social lives affect our brain health, we should turn our attention to claims of a huge looming threat to public health – a loneliness epidemic. In 2018, Fox News reported on 'America's Growing Loneliness', and there are hundreds if not thousands of media reports all making similar claims, published by Forbes, *Harper's Bazaar*, Yahoo, the *New York Times*, the *Daily Mail*, *The Times*, the BBC, France24 and many, many others. To read the headlines in the media, one could be forgiven for believing that this epidemic is tearing through modern society.

What is really going on? As we go through life, do we become more lonely? Are people lonelier today than at other times? A rigorous review of all the available data clearly shows that the perhaps surprising answer to the first question is 'no': we do not become more lonely as we age. Rather, the reverse is true – we become less lonely. But the relationship between age and loneliness is quite a complex one. Data from the UK Office of National Statistics show that if you are aged 16–34 you are more likely to report feeling often or always lonely (both 8 per cent) than 50–64-year-olds (5 per cent). Moreover, a recent review by the American researcher Louise Hawkley showed that reported loneliness continues to decline up to the age of 75, after which it starts to increase. How do we explain these trends? It appears that after 50, we become more socially mature and adapt our behaviour and expectations, so muting our experience of loneliness. After 75, ill-health and the loss of spouses and partners starts to take its toll.

In answer to the second question, again there is, surprisingly, little evidence that we are lonelier now than in previous generations – regardless of age. A US study published in 2019 found that people born between 1948 and 1965 were no more likely to feel lonely than those born between 1920 and 1947, and that those older adults had not become lonelier over the decade from 2005 to 2016. A number of cross-sectional studies in Sweden looking at people aged 85, 90 and 95 also showed no increase in reported loneliness over a ten-year period – and the same finding applies to studies looking at younger people. Two important US studies looking at college and high-school graduates, one over a thirty-year period between 1976 and 2006, and one analysing two periods (1978–2009 and 1991–2012), both found that these young people were no more likely to report feelings of loneliness than their counterparts in earlier generations.

The widespread use of mobile telephones, the internet and social media may provide part of the explanation for why younger age groups are more likely to say that they are lonely than older people. If we look at the evidence, social media is not so social. To quote one study among many, a survey of nearly two thousand young people aged 19–32 by a team from the University of Pittsburgh, 'Young adults with high social media use seem to feel more socially isolated than their counterparts with lower use.'[1] To be precise, it was found that those who use social media frequently are more than three times as likely to feel lonely as those who don't. In many ways, our hand-held devices keep us self-absorbed in situations which otherwise would provide a basis for casual interaction and conversation with those around us. They keep our heads down, our eye contact with others subdued and our deep-seated human need for social interaction suppressed. Furthermore, social media use has been linked to the phenomenon of FOMO, or 'fear of missing out', a term popularized by Patrick J. McGinnis in 2004, four years after it had first appeared in an academic journal. FOMO has been shown to lead to loneliness and, unsurprisingly, lower self-esteem.

The media headlines asserting a loneliness epidemic are largely erroneous, many of them reporting single studies and based on the flawed reasoning that living alone automatically means you will be more lonely. In fact, spending time alone is not a good predictor of whether you will feel lonely or whether you have good social support. Numerous studies support this conclusion, including a large longitudinal study over two years (2006–8) in Boston, involving twelve thousand older people. The researchers found that there was almost no correlation between being alone and feeling lonely – but they did find that living by yourself was associated with poor physical health, and that feeling lonely was associated with poorer mental health.

The truth of it is – *there is no new loneliness epidemic*. Loneliness is part of the human condition, and it is entirely to be expected that it should be widely experienced. Not just evolutionary scientists but ancient philosophers such as Aristotle and Plato and their more modern successors, including Martin Buber and Jean-Paul Sartre, have all declared loneliness to be an essential element of human existence. Poets across the ages have written evocatively of loneliness – for example, Milton in *Paradise Lost*. As Satan steps into the Void, he declares: 'From them I go / This uncouth errand sole, and one for all / Myself expose, with lonely steps to tread / Th' unfounded deep.'[2] And centuries later, Albert Camus conveyed the profound and paradoxical impact of loneliness almost shockingly in his 1942 book *The Stranger*, in which the subject finds relief from his loneliness, even happiness, in the anticipation of a crowd filled with hate:

> For the first time, in that night alive with signs and stars, I opened myself to the gentle indifference of the world. Finding it so much like myself – so like a brother, really – I felt that I had been happy and that

I was happy again. For everything to be consummated, for me to feel less alone, I had only to wish that there be a large crowd of spectators the day of my execution and that they greet me with cries of hate.

Nor is a concern with loneliness the preserve of 'high' culture. Nothing better illustrates its centrality to our common experience than the number of popular songs about loneliness: some 45,000 since 1950, including gems like 'Only the Lonely' (Roy Orbison), 'Lonely This Christmas' (Elvis Presley) and 'Eleanor Rigby' (The Beatles).

So being alone, and/or lonely, is not unusual; but it can nevertheless have severe consequences for our general health, and specifically for the health of our brain.

Social connectedness and health

The Norwegian rat is a wonderful animal. Furry and cute, it is naturally gregarious. It spends significant time in physical contact with other rats. It burrows extensively and forms complex social relationships. It cooperates with others in raising its young. In fact, it's a quintessentially social animal. And, like many animal populations, it suffers a naturally occurring incidence of cancers – spontaneously developing both benign and malignant tumours. Researchers in Chicago randomly assigned a sample of Norwegian rats to live either alone or in groups, each group made up of five female rats. They made an astonishing discovery. Such was the effect of social isolation that the incidence of mammary tumours in the isolated rats increased to *84 times* that of the group controls, and malignancy *by a factor of three*, specifically for two types of cancer – which, interestingly, are the same as the breast cancers in women most commonly identified in the initial stages. Unsurprisingly, the isolated females showed measurably higher stress levels. They also showed noticeable and significant changes in behaviour – they were more anxious, more fearful and more vigilant.

Numerous and rigorous scientific studies over the past twenty years have shown equally clearly the pernicious effects of social isolation and loneliness in humans. Eloquently making a distinction between the two, in an article in the *New Yorker* published as the Covid-19 pandemic was accelerating in spring 2020, Jill Lepore directed us away from our obsession with the perils of infectious disease:

> You can live alone without being lonely, and you can be lonely without living alone, but the two are closely tied together, which makes lockdowns, sheltering in place, that much harder to bear. Loneliness, it seems unnecessary to say, is terrible for your health.[3]

And indeed she is right. We have to make a distinction between the two, because the evidence shows that they act independently on health; but neither social isolation nor loneliness does us any good. Social isolation, quite simply, is the absence of human contact. It includes staying at home for lengthy periods, having little or no communication with family, acquaintances or friends, and avoiding contact with other humans when such opportunities arise. Sounds familiar? Researchers have grasped the opportunity presented by the lockdowns introduced in response to the Covid-19 epidemic, and there are currently long-term data collection studies going on in the UK (at University College London), in Australia (at Deakin University) and in the USA (at MIT's Saxelab).

We already know enough to say with some certainty what the effects of social isolation will be, enforced or otherwise. In a 2012 study at the University of California, Berkeley, Dr Matt Pantell and his colleagues analysed data from a large sample of some twenty thousand adults aged between 17 and 89. They found that social isolation was as powerful a predictor of mortality as conventional clinical risk factors, such as cigarette smoking and high blood pressure. In fact, social isolation was a bigger predictor than high blood pressure and just as big as smoking. These findings were corroborated by an even larger meta-analysis carried out in 2014 at Brigham Young University. This kind of study makes calculations from the combined results of hundreds of studies worldwide, in this case with data from over 3.4 million people. Again, it yielded astonishing results: the increased risk of mortality from social isolation was found to be 29 per cent; from loneliness, 26 per cent; and from living alone, 32 per cent – increases of the same size as those attributable to the conventional factors contributing to mortality noted above. Other researchers have made equally startling statements – for example, that lack of social connectedness is as lethal as smoking fifteen cigarettes or drinking a bottle of gin a day, or being morbidly obese. Different studies have shown that, regardless of age, socially isolated people show high relative risks of musculoskeletal disorders, moderate to severe depression and multiple general health problems, and are more likely than others to define their own health as poor. These health consequences may be explained in part by the tendency of social isolation to lead to unhealthily low levels of physical activity, poor diet and use of psychotropic medicines.

In research, loneliness is generally defined as 'feeling lonely more than once a week'. There are two preconditions for loneliness: a lack of meaning in our relationships; and a sense of ourselves as separate from others. As social primates, we are wired for intimacy, and unresolved loneliness has drastic consequences. Loneliness is related to a number of health conditions, but the big giveaway is its impact on mortality: the data tell us that people who are not lonely have a 50 per cent better chance of survival than those who are lonely – or, to put that in slightly more technical terms, people who are not situationally lonely are 50 per cent less at risk of dying than those who are. There is a long list of health conditions related to chronic loneliness; these are summarized in box 6.1.

What are the reasons for these deadly effects? Generally, those who are socially isolated from friends, family or neighbours over many years have poor social skills, lack social support, have less opportunity to access health care and are at greater risk of loneliness. In addition, these conditions lead to important behavioural changes – such individuals have been found to have high anxiety levels, to regard interpersonal relationships as less pleasant than other people, to be on the look-out for social threats, to be less capable of dealing with stressful situations and to have poor health-related behaviour, such as high levels of smoking and alcohol consumption.

BOX 6.1: HEALTH EFFECTS OF SOCIAL ISOLATION AND LONELINESS

- Social isolation increases the risk of mortality by about 30 per cent.
- Social isolation is a powerful predictor of mortality, as much as high blood pressure and tobacco smoking.
- Regardless of age, socially isolated people show high relative risks of poor self-reported health, musculoskeletal disorders, moderate to severe depression and multiple general health problems.
- Social isolation often leads to unhealthily low levels of physical activity, poor diet and use of psychotropic medicines. In virtually all research where it is measured, social isolation correlates highly with low physical activity and is a determinant of it.
- People who are not lonely have a 50 per cent better chance of survival than those who are.
- Severe loneliness at all ages contributes to depression, alcoholism, thoughts of self-harm, aggression and suicide, and to actual suicide.
- Loneliness exacerbates the pain, depression and fatigue which often accompany serious long-term illness.
- Loneliness is associated with an increased prevalence of high blood pressure, high cholesterol and obesity.
- Loneliness raises the levels of the stress hormone cortisol. Sustained high levels of cortisol are related to anxiety, depression, digestive problems, sleep problems and impaired immunity.
- Chronic loneliness is related to an increased risk of stroke, heart disease and some types of cancer.

But there are also biological explanations – and unsurprisingly, in this age, when we know so much more about our genome, DNA features prominently. There isn't the evidence to say there is a specific 'loneliness gene', but in a BioBank study in the UK drawing on some 487,647 participants, the scientists identified *fifteen gene regions linked to loneliness*. The evidence was clear: the differences between people's responses to being alone – some find it welcome, others find it excruciatingly painful – may well be determined by our DNA. And further, the same 'loneliness' genes appeared to be linked to obesity, suggesting that the two conditions might be associated. There were also genetic overlaps with traits identified in previous studies: depression, obesity and, in particular, poor cardiovascular health. Another explanation is that chronic loneliness accelerates our ageing process. Lonely people have been found to have shorter telomeres at the ends of their chromosomes. As noted in an earlier chapter, shorter telomeres are associated with many age-related diseases and with early mortality.

In summary, both social isolation and loneliness affect the body – but in slightly different ways which we do not yet understand completely. As we saw in chapter 1, inflammation is the body's response to stress. It signals the immune system to heal and repair damaged tissue; but if it is prolonged it eventually starts to damage healthy cells, tissues and organs, giving rise to chronic long-term illness, such as heart disease, stroke, and lung and kidney disease. Clearly, social isolation is a powerful stressor, and so will raise levels of inflammation. Recent research in the UK shows that social isolation raised levels of two proteins, CRP and fibrinogen, that are inflammatory markers in the blood, and that – for reasons not yet understood – this effect was stronger in men than in women. The physiological effect of loneliness was different. It has been associated with a different inflammatory marker called Inter-Leukin 6 (IL6). These differences reinforce the point made earlier that being alone (socially isolated) and being lonely are not the same thing. But do these findings throw any light on how isolation and loneliness affect the brain?

Social connectedness and the brain

In mid-September 2018, two experienced poker players sat across a table in Las Vegas and made a very unusual wager. If one of them could remain in totally blacked-out isolation for thirty days, the other would pay him $100,000. As Ray Alati stepped into the room, he was confident of winning. Three days later, the hallucinations started. And the foolhardy experiment ended short of the target date by a full ten days (for those curious among you, Alati negotiated an early pay-out of $62,000). His fellow player didn't even try it.

The world record for the longest spent in isolation is neither voluntary nor lucrative. And there are several candidates for the award – all of them prisoners of the justice system. In the USA, the contender is Albert Woodfox, one of the notorious 'Angola Three', held in solitary confinement for some forty-three years. In the UK, it is Robert Maudsley, still imprisoned aged 64 in his thirty-ninth year of solitary. (For readers not familiar with Maudsley's case, he killed three fellow inmates and ate the brain of one of them.) But this is not the place to discuss the rights and wrongs of solitary confinement. These unfortunate examples serve as some kind of abhorrent natural experiment, in which the effects on the brain of such prolonged isolation can be observed – not over a mere thirty days but over thirty or more *years*.

The USA has the dubious honour of having the oldest records in the history of research on solitary confinement, dating back to 1829, when medical observers at the Eastern State Penitentiary in Philadelphia noted the mental and physical disorders among its solitary inmates. This did not escape the attention of Charles Dickens, who visited the Philadelphia Penitentiary and described solitary confinement as the 'slow and daily tampering with the mysteries of the brain . . . immeasurably worse than any torture of the body'.[4]

And so modern science has proved. While each individual's experience is different, there is a long, long list of behavioural and psychological

Solitary confinement

effects known to result from solitary confinement: hallucinations, increased anxiety and nervousness, feelings of being watched, diminished impulse control, severe and chronic depression, appetite loss and weight loss, heart palpitations, talking to oneself, problems sleeping, nightmares, self-mutilation, difficulties with thinking, concentration and memory, and lower levels of brain function.

All these effects on the brain and the mind (which we may define as the brain's product in consciousness – our thoughts and feelings) inevitably feed into the body through a powerful mind–body pathway which physiologists call the 'stress system'. The stress system is the ghost in the machine – a relentless, primeval, self-regulating physiological mechanism, millions of years old, residing in the brain to protect it against threats or 'stressors'. Its role is so important that similar systems are found in virtually every animal on the planet. Stressors can be physical or psychological, and the two kinds of stressor are processed by different but overlapping brain circuits. These circuits are exceptionally complex and a full description of them lies beyond the scope of this book. For our purposes here, a simplified summary will serve. Physical stressors (such as pain or cold) are given top priority by the nervous and circulatory systems, which respond using two pathways. The first one (the sympathetic adreno-medullar system or

SAM) is concerned with alertness, vigilance, situational awareness and decision-making, and instigates rapid, brief responses. The second one (the hypothalamic–pituitary–adrenal or HPA axis) works much more slowly and is designed to activate a sustained stress response through the release of the stress hormones called corticosteroids. These two pathways influence many functions of the body, including digestion, the immune system, mood and emotions, sexuality, and energy storage and expenditure. Psychological stressors (such as uncontrollable events or an unsatisfied internal drive) not only activate their own set of circuits but trigger the SAM and HPA responses *as well*. Those additional circuits tap into powerful components of the brain, including the amygdala and the hippocampus, parts of the limbic system which, as we have already seen, is the 'emotional seat' of the brain.

The point of this brief description is that if we become socially isolated or lonely, both sets of stress responses kick in. We experience a profound and substantial 'neuro-endocrine' reaction which has deep and far-reaching consequences. Even the genetic mechanisms controlling the chemistry of these responses are implicated. And science has got to the point where, astonishingly, we now understand these molecular changes – with what seems an unthinkable result: a pill for loneliness.

Social isolation – messing with the brain

The humble, peace-loving mouse is not the creature you think it is. Isolate it for a mere fourteen days and it cannot be placed back with other mice, for fear of the savage, unprovoked attacks it will launch on them. And the blame for this ferocious anti-social behaviour? It lies deep in the chemistry of the brain.

A vivid picture of what transpires in the brain of a more or less tranquil mouse isolated from others has recently been exposed for the first time. The bad behaviour of the isolated mouse has long been known – hyperaggressive on contact with others, while fearful and oversensitive to external threats, reflected in behaviour such as 'freezing' repeatedly.

Now animal scientists at the California Institute of Technology have found exactly what is going on. In an isolated mouse, certain areas of the brain such as the amygdala and the hypothalamus become saturated with a powerful brain molecule called 'neurokinin beta' (NKB for short). NKB is produced by a gene called Tac2 and then binds to other brain cells, changing their behaviour – and, more to the point, the activity of their neural circuits. If the scientists 'switched off' Tac2 in the amygdala, fear diminished but aggression rose. In the hypothalamus, the reverse happened – aggression went down, fear went up. It appears that the release of such compounds right across the brain coordinates the changed and inconsistent behaviour of the isolated mouse.

Nor are these the only changes in brain chemistry brought about by isolation. Scientists at MIT and Imperial College London used a 'three-chamber test' in which a socially isolated mouse is placed in the middle part of a box with a chamber either side. It can choose to go left or right. If another mouse is present in one of these chambers (the 'social chamber'), the isolated mouse makes a beeline for it. Non-isolated mice will choose either chamber randomly, and if they choose the social chamber will generally ignore the other mouse. But this is not the point of the story. These researchers were looking at a particular part of the brain called the dorsal raphe nucleus (DRN), which produces dopamine, the 'reward' hormone. If they put a non-isolated mouse in the central chamber and turned on its dopamine cells using a light pulse, it spent more time in the 'social' chamber, with a second mouse. If they put an isolated mouse in the central chamber and *switched off* its dopamine cells, the expected sociability was not restored. These findings suggest that dopamine cells, when activated, motivate sociability in response to the negative and aversive state of isolation. The drive for company in isolated mice is thought to act like hunger (or any other compulsive appetite). Different neural circuits are activated when the mouse wants social interaction because it is rewarding (like delicious food) and when it feels lonely (as when it's hungry).

Dopamine isn't the only prompt for such behaviour. A group of scientists at Toronto University also looked at the mouse DRN, this time focusing on the DRN cells that produce serotonin, another brain hormone. They discovered that social isolation reduces the secretion of serotonin – and they discovered how to block the action of social isolation by administering a substance which restored serotonin secretion. By doing so, they found that they could counter the negative behaviour in their isolated mice (in this case, eating disorders and lack of mobility caused by anxiety). The scientists in these studies made a hugely significant claim – that they had discovered the 'loneliness centre' of the brain, the DRN.

People are not mice. And for ethical reasons we cannot perform these experiments on humans. But they shine a ray of light into a dark corner of our brain health. These fascinating insights show us that loneliness is controlled to a large extent by the chemistry of the brain. And though loneliness is unpleasant, it will not kill us. It is our brain telling us that something is seriously wrong. It drives us to change our behaviour – compelling us to seek out others and optimize our social contacts. However, like unrequited hunger, if we do not satisfy it, loneliness will lead to damaging consequences – anxiety, nervousness, fear, depression. And it will even change the structure of our brain.

The incredible shrinking brain

It's pretty much odds on that if you are often lonely, the structure of your brain will change as a result. Your brain will in all probability be very different from that of a comparable non-lonely person. It is only relatively recently that we have been able to make such a statement. Until the last few years, we knew – or suspected – very little about the effects of loneliness on brain structure. Now the veil is being lifted. And what we see is far from pleasant.

Researchers in Berlin measured how lonely people were and then took images of their brains. They found that certain areas of the brain were smaller in those who were the most lonely. Which areas were these? The answer should not surprise us. There were two regions, both deep in the brain and both involved in regulating our emotions – the amygdala and the hypothalamus. These results did not surprise the researchers either. Previous studies had already shown that the size of the amygdala is related to how social we are – bigger social network, bigger amygdala. Two points need to be made straight away. First, the results didn't show cause and effect; and second, we don't have an explanation. But we know enough to make some informed speculations. It could be that the release of stress hormones (such as corticosteroids) during prolonged loneliness raises blood pressure and so damages the hypothalamus. Or it may be that there is an interaction between loneliness and ageing – both of which are associated with a smaller amygdala.

There is other evidence, too. Researchers at University College London looked at a different part of the brain – a particular region on the left side which is known to be involved in basic social activity (the posterior superior temporal sulcus or PSTS for those of you who are technically minded). They got the same result: lonely individuals have a smaller sulcus. These individuals also had difficulty responding to social cues, an ability vital to forming and maintaining relationships with others. Conversely, in another study of 250 people living at home, enriching the participants' social lives not only improved their thinking skills but after forty weeks resulted in significant increases in brain volume compared to a control group.

How do we explain such changes in brain volume? And can they be prevented or reversed? It's been found that when mice are moved from a rich, playful environment to complete isolation, the overall size of individual neurons in the brain *shrinks by 20 per cent after about a month*. At first, moreover, the cells start to grow many more connections – as if the brain was trying to 'save itself'; but after three months, the new connections recede, as if the brain had 'given up'. On top of this, there's a steep decline in levels of that important brain protein BDNF, which as we've already seen plays a critical role in the growth and survival of new brain cells. Unsurprisingly, stress hormones such as the glucocorticoids were high, and the isolated mouse brains had more fragmented (broken)

DNA in their cells.

There is one final question to examine. And it is one of a very sensitive and compelling nature. So far we have been talking about the mature brain. Brains that have grown up. What happens to the young brain, to the brains of children who are subjected to social deprivation and isolation? There is a wealth of social evidence to show beyond question the appalling consequences of this kind of deprived childhood. And zoologists will happily tell us that social animals don't just prefer contact with others – they *need* it if they are to develop properly. But do we know what happens to brain development if we are socially deprived in childhood? The development of the tracts of white matter in the brain – which enable normal communication and processing by the grey matter – is central to the effects of social neglect in children. The tragedy of the Romanian children incarcerated in orphanages in the Ceaușescu era of the 1980s provides ample evidence. An imaging study of a small number of socially deprived children adopted into the USA from state institutions in Romania showed that they had far less white matter than normal children, particularly in a structure called the 'uncinate fasciculus', which connects parts of the brain's temporal lobe, including the amygdala, with parts of the pre-frontal cortex – all areas vital to emotional growth and social development. This evidence, though, while undeniably interesting, does not tell us *how* poor social conditions can produce these changes – or the extent of their role in determining social outcomes. For this, we have to turn yet again to animal experiments.

In one such experiment, four mice were separated and kept alone, and another group was housed normally, four mice per cage. After four weeks the social behaviour and memory of the isolated mice were worse than those of the group mice. After six weeks, the researchers inspected some of the brain cells in the pre-frontal cortex of the mice. Sure enough, although the *number* of cells was the same in both groups, the isolated mice had stunted support cells (oligodendrocytes) with simpler shapes, fewer branches and so on, and two genes that produced important proteins were turned on ('expressed') less often in their pre-frontal cortices.

Electron microscopy also revealed that the isolated mice had undergone significant thinning of the myelin sheaths – the insulation around nerve fibres that is important for passing messages in the brain. It was further identified that there was a critically sensitive period – the fourth and fifth weeks of the experimental period – in which the damage was established. Without going into the detail of this extremely sophisticated and clever piece of science, we can say that the 'blame' for that damage was found to lie with certain vital genes that were not expressed. Of course, no one is suggesting that the effects of social isolation are confined to white matter damage. But this evidence shows the subtle cellular mechanisms by which social isolation affects normal brain development.

The main effects of social isolation on the brain are summarized in box 6.2.

BOX 6.2: SOCIAL ISOLATION AND THE BRAIN

- All social animals, including humans, fail to thrive if socially isolated, and suffer multiple harmful consequences, physical and behavioural.

- Persistent social isolation is stressful, increases inflammation, suppresses the immune system, and provokes harmful hormonal changes which affect the brain.

- Social isolation drives powerful chemical responses in the brain, involving powerful and primitive structures such as the amygdala and hypothalamus.

- Two principal chemical responses involve the release of two messenger molecules in the brain – dopamine (the reward hormone) and serotonin (which elevates mood).

- Loneliness is associated with changes in brain structure: smaller volume in certain brain areas, smaller individual brain cells and reduced connections between cells, including the white matter tracts.

- A rich social environment is necessary for the normal development of the brain in children.

Social connectedness and brain health

Ethical constraints mean that scientists studying brain health and social engagement in humans have largely relied on epidemiology – studies that are purely observational. These studies, which often involve very large numbers of participants, compare people on the basis of their different types and levels of social behaviour, rather than manipulating their level of social connectedness (as we can do in animal experiments). In this way, the impact of social engagement on people's brain health is assessed.

Social engagement is very often mixed up with other activities that may also influence brain health, such as cognitive stimulation (for example in playing a board game) and physical activities that occur as part of our social life. It is therefore difficult to tease out which activity is responsible for which outcome. Is it the purely social part of an activity, or the physical or mental challenge, or even the combination of them? Though observational studies cannot conclusively establish that social engagement directly causes improvements in or maintains brain health, the weight of the currently available evidence strongly suggests that good social engagement *is* linked to better brain health. Current expert opinion recommends that we engage in active social participation throughout our lives and stresses that there are other potential health benefits if we are socially active. We know that becoming more socially active has few, if any, harmful effects, and also that those of us with better brain health tend to seek out more frequent and higher-quality social engagements – a truly virtuous circle.

The world's foremost impartial authority on the subject, the Global Council on Brain Health, agrees that the evidence shows a positive impact of social engagement on brain health, including benefits to our thinking and reasoning abilities. However, they rightly argue that more work needs to be done to give us more precise guidance, by improving the growing body of research evaluating the various effects of different aspects of social engagement on memory and reasoning skills.

For some people it happens slowly, for others fast; for some it comes early, for others, late. But we all lose some of our thinking skills as we age. And there is an interesting dilemma which scientists have struggled to resolve – when we reach older age, does chronic loneliness cause cognitive decline, or is it a consequence of it? Resolution came in 2017, when Harvard University scientists published the results of a twelve-year-long study. In 1998, they recruited 8,311 adults aged 65 or older from the US Health and Retirement Study. Participants' social networks were assessed by noting whether they were married or with a partner, had contact with friends or neighbours at least weekly, had contact with a child at least weekly and/or participated in volunteer activities. They were assessed for depression, loneliness and thinking skills (cognitive function).

What did the researchers find? In 2010 the study was completed and the results analysed. Some 18 per cent of the participants were described as lonely at the start, and this loneliness was associated with a *20 per cent faster rate of cognitive decline over twelve years compared to participants who were not lonely.* This finding was independent of their social class, their financial status, their social network, their health and 'baseline' depression (i.e. at the start of the twelve-year study). In other words, loneliness was a strong predictor of cognitive decline, independently of these factors. Interestingly, depression over time was related to faster cognitive decline as a factor in its own right.

In 2019, a meta-analysis (a mathematical review of multiple studies) looked at the relationship between loneliness, mild cognitive impairment and dementia. It covered only ten studies, but each of those studies included large numbers of people – in total, some 37,000 aged 65 to 83 years. The results for dementia were striking: loneliness was a strong risk factor. There was more limited evidence for mild cognitive impairment, but the results still suggested that loneliness had a contributory effect.

A lot of us, however, will want to know what happens from our earlier years – in our twenties, thirties and forties. For that, we have to turn to longitudinal or 'lifelong' studies. Over fifty of these studies were collectively analysed by researchers at Exeter University in 2018. Though these various studies used many different indicators of social isolation and social activity, the findings were again striking. The extensive analysis found that, across our lifetime:

- engaging in social activity is significantly associated with better cognitive outcomes;

- larger social networks are significantly associated with better cognitive function;

- when looked at together, social activity and social networks are significantly associated with thinking skills;

- where men and women were looked at separately, the effect of larger social relationships was similar for men and women, with a slightly greater effect for women.

Slade School of Art Women's Life Class

The researchers concluded that 'aspects of social isolation, including low levels of social activity and poor social networks, are significantly associated with poor cognitive function in later life'.[5] They also concluded, in line with many other studies, that poorer social relationships throughout life increase the risk of dementia.

The key points about social connectedness and brain health are summarized in box 6.3.

BOX 6.3: SOCIAL CONNECTEDNESS AND BRAIN HEALTH

- At all ages, a good social life is linked to better brain health.
- Social engagement helps to maintain thinking skills throughout life.
- Good social connectedness slows cognitive decline in later life, notably after middle age, and individuals who are more socially engaged have a lower risk of cognitive decline.
- Engaging in social activity and having larger social networks is associated with better cognitive outcomes in later life.
- Loneliness is a strong risk factor for dementia, which develops through many years of slow neurodegeneration.
- Social contact can help to improve memory formation and recall, and protects the brain from neurodegenerative diseases.
- Building up good cognitive reserve through a rich social life, education and learning new skills at any age is thought to protect brain health.

Staying connected

If there is one single overriding message from the research in this chapter, it's that throughout our lives, our 'social capital' really matters, if we wish to preserve our brain health. By 'social capital' we mean engaging in social activity and maintaining good social networks. This means cultivating a lifestyle that both helps us to deal with being alone and reduces our feelings of loneliness by enabling us to be with others. Does the research tell us how to go about this?

The first principle has to be: 'Start with yourself.' Research has shown that a positive 'frame of mind' is hugely valuable – it stimulates good health, influences the immune system and actually prolongs healthy life. In her famous paper of 2002, Beccy Levy of Yale University showed that positive thinking can lengthen our lives by more than seven years. And other research has shown that negative emotions are damaging to both our physical and our mental health. *So, dispelling negative thoughts about being alone is a great start.* And it is one of the few approaches that has been shown to work. Psychologists call this 'changing maladjusted thinking'. In a recent study in Chicago, it was found to be four times more effective than any other approach. We can remind ourselves that everyone becomes lonely at some point, and we are not unusual in feeling alone. It's also important not to criticize or blame ourselves for being alone. In fact, feelings of loneliness are instrumental in motivating us to seek the company of others. Self-perception is immensely important in dealing with loneliness. Researchers at Cornell University ran a series of experiments in which two people met and talked for the first time, then rated their own conversation and the other person's. Across conversations of all lengths, some with topics provided and some not, the researchers found that people consistently rated their conversation partner as more likeable and enjoyable to talk to than themselves. The message is clear: generally, other people like us better than we think – which is another reason why we should not fear approaching others.

The next principle is: 'Little things make a difference.' There's a lot of stigma attached to loneliness in modern society, and research has shown that many people hesitate to take even the simplest steps to talk to or reach out to others to reduce their feelings of loneliness. Some interesting work by Gillian Sandstrom at Essex University showed that we have forgotten how to talk to those we don't know – passing conversations during our commute, while out shopping or walking the dog, or at the school gate. But it also produced two other very helpful findings: first, that we can relearn this lost art by having many small interactions with others; and second, that just talking to strangers and acquaintances not only alleviates loneliness but gives us a more positive mood overall and a better sense of being connected. There is huge benefit in brief snippets of conversation and the common decencies of salutation. Saying 'hello', 'good morning', 'how are you?' may seem trivial. But such brief greetings are hugely beneficial in alleviating loneliness. Share a smile a day with someone, hold a door open, practise a random act of kindness. Reach out to neighbours or acquaintances to whom you may not have spoken in a long time: ring up, send a card or email, or check for contacts on social media. The power and usefulness of this principle – which some have called 'the surprising power of weak ties' – are reflected in the number of current campaigns and initiatives in public life on the same theme – 'Britain Get Talking' on ITV, 'Crossing Divides' on BBC TV and the UK National Rail campaign, 'On the Move' – all designed to end the erosion of conversation in modern life.

Third: 'Think of others.' Overturning decades of received wisdom based on Freud, Maslow and other, more modern, psychologists, Dr Michael Babula argues that we have hitherto followed a flawed premise in assuming that we humans are inherently self-interested, and that the drive to maximize our self-interest is a healthy guiding principle in life. By contrast, he argues that 'it is this drive towards self-interest and materialism, at the expense of our wider social surroundings, that is a major source of crime, cruelty, unhappiness and lack of empathy within western society'.[6] In thinking of others, we reduce our own pain of loneliness as we seek to be with and help our fellow humans. And in doing so, we break the bonds of social isolation: thus, paradoxically, by seeking to help others, we help ourselves.

And that neatly brings us to the next principle: 'It's the group that matters.' Of all the findings presented in this chapter, the most fundamental is that as humans we are inextricably wedded to the group. We are social beings. Without social contact, it's not just our physical health and our well-being that suffer; our brain health suffers too. Being a member of a group, formal or informal, gives us a sense of purpose, satisfies our need to belong and gives us the relationships we need to thrive. And, more importantly, good lifelong social capital contributes to our brain health, from childhood to our most senior years. It is therefore important to be proactive from our earliest adult lives in building and maintaining our social networks, and to be socially active. This may involve maintaining friendships, staying in touch with family and distant relatives, catching up with old friends, or being a member of social groups or clubs. One exceptionally beneficial activity is volunteering, for which there are endless opportunities, especially through charities or public bodies such as art galleries, museums, heritage organizations, environmental groups and so on. The internet and social media may be helpful in making contacts, but they do come with a downside. Some, indeed, have gone so far as to say that the internet and social media are huge drivers of loneliness, allowing us to remain in touch with each other remotely without actually having to connect with people in actual time and space. I referred earlier in this chapter to some research showing that those who spent more time looking at their screens and used the internet more heavily were more prone to loneliness. Among the possible reasons for this is that such habits may hinder the development of good social skills. They may also get in the way if you are in company, or may make you miss the vital social cues on which interpersonal ties are founded.

There's one particular burgeoning social use of the internet that we need to consider: dating websites. When they were launched in 1995, the idea that singles could meet one another on the internet was revolutionary, and generally met with scepticism, even derision. And yet online dating is now the second most widely used paid-for online service. It is estimated that about 20 per cent of all marriages result from dating websites and that 'online' marriages occur more rapidly, last longer and are less likely to end in divorce. But it's not all good news. Studies have shown that use of these sites, such as Tinder (which gets 1.8 billion swipes per day), can lead to addictive behaviour because their unpredictable reward

Hip, Hip Hurrah! Artists' Party at Skagen

ratio hijacks the reward system of the brain (via dopamine release). And further, studies have shown they can lead to lower self-esteem and to depression – not very promising outcomes for good brain health.

And so to our last principle: 'Enrich your life.' Research has shown that simply reducing social isolation is necessary but insufficient to maintain brain health. It must happen alongside other lifestyle factors, such as a healthy diet, ample exercise, good sleep and so on. It is also extremely important to build a rich cultural life and to make the most of our free time. Enjoy music, art and theatre. Appreciate and experience the natural environment. Read books, go to talks, listen to the radio and watch television. Take the opportunity to extend your education or to learn new skills. All these activities go towards building what psychologists call 'cognitive reserve' – the mind's resistance to damage of the brain. We can think of cognitive reserve as our brain's ability to improvise and find alternate ways of surviving critical tests. Just as a powerful engine enables us to accelerate a vehicle around an obstacle, so our brain can be encouraged to make additional resources available to cope with sudden challenges. Cognitive reserve is developed by a lifetime of education and experience to help our brain cope better with failure and setbacks. So it should come as no surprise that a combination of lifestyle factors throughout our lives boosts cognitive reserve and helps to maintain a healthy brain.

These five key principles of 'staying connected' are summarized in box 6.4.

BOX 6.4: STAYING CONNECTED – THE FIVE KEY PRINCIPLES

1. *Start with yourself.* Dispel negative thinking and be positive about yourself; challenge negative thoughts and recognize that loneliness is a natural part of being human; understand that everyone at some time feels lonely and just like you.

2. *Little things make a difference.* Relearn how to talk to others, particularly those you don't know; make small talk in passing conversations, starting with simple greetings such as 'Good morning' or 'How are you?' Reach out to others, such as neighbours or distant friends or even family. Be prepared for negative responses from others and don't be put off by them.

3. *Think of others.* Be less selfish and more altruistic; develop empathy. Be less self-centred and think more of the needs of others; try to help those who are less fortunate or in need; engage in charitable giving or activity.

4. *It's the group that matters.* Be proactive in building and maintaining your social networks. Be socially active; look for group activities which generate a sense of belonging. Try volunteering and working in a group. Go online to make contact with others, but not as a substitute for real social contact.

5. *Enrich your life.* Make the most of the time you spend alone. Build a rich cultural life, with others if possible; use cultural activities as a means to share experiences. Build up your 'cognitive reserve'.

The unthinkable pill?

Finally, let's contemplate the unthinkable. It seems there's a pill for pretty much everything that ails us psychologically, including anxiety and depression, so why not for loneliness? One of the candidate substances is a molecule that occurs naturally in the body. This is pregnenolone, a steroid made in the adrenal glands from cholesterol. It has been shown that many of the effects of social isolation in small mammals can be attenuated by a single injection of this hormone. Originally prescribed for female health as a precursor molecule for oestrogen, it is now used at the discretion of physicians alongside testosterone and oestrogen to help relieve symptoms of oestrogen imbalance and menopause in women, and andropause and testosterone deficiency in men. In the USA it is freely available over the counter as a dietary supplement. It is said to boost mood, enhance well-being, and improve sleep, memory and thinking skills, helping to increase the growth of neurons in the brain and healthy white matter. It is currently under clinical trial in the USA for pharmaceutical use. However, a small caveat: hardly any of the experts involved in these trials use this therapy as anything more than an 'adjunct' to other, proven ways of mitigating loneliness.

Some of the recommendations from the Global Council on Brain Health, addressing the needs of people over the age of 50, are shown in box 6.5.

BOX 6.5: PRACTICAL WAYS TO STAY CONNECTED – ADVICE FROM THE GLOBAL COUNCIL ON BRAIN HEALTH

1. Focus on the relationships or social activities you enjoy the most. Be active and challenge yourself to try out organized clubs, courses, interest groups, political organizations, religious gatherings or cooking classes.

2. If there are barriers to interacting with people (e.g. difficulty getting around, unsafe neighbourhood), see if you can identify

someone you could ask for help, perhaps through a charity or helpline, and let someone assist you in making connections.

3. Try to keep a circle of friends, family and/or neighbours with whom you can exchange ideas, thoughts, concerns and practical matters, and who can also help or encourage you.

4. Try to have at least one trustworthy and reliable confidant to communicate with regularly – say, every week: someone you feel you can trust and count on.

5. If you are married, this can benefit your cognitive health, but you should consider fostering other important relationships too.

6. Try to communicate every now and then (perhaps monthly) with relatives, friends and/or neighbours – whether in person or by phone, email or other means.

7. Get into the habit of helping others, whether informally or through organizations or volunteer opportunities. For example, visit a lonely neighbour or friend, shop for/with them, or try cooking together.

8. Maintain social connections with people of different ages, including younger people. Keep in touch with grandchildren, or volunteer to help people at a local school or community centre.

9. Think about the skills you have and that you use routinely that might be worth passing on to others. Offer to help teach a younger person skills you may already have, such as cooking, organizing an event, assembling furniture, saving for the future, investing in the stock market etc.

10. Add a new relationship or social activity you haven't tried before. Place yourself in everyday contexts where you can meet and interact with others (e.g. shops or parks).

As a postscript, it has to be acknowledged that all this counsel flies in the face of the global public health advice on 'social distancing' given during the coronavirus pandemic, as a countermeasure to the high risk of contagion and the potential severity of the disease. No one within

2 metres of anyone else. No social gatherings. No visits to the old or vulnerable. No sporting events. We don't yet know the longer-term effects of 'lockdown' in terms of social, psychological or cognitive health, but one thing is starkly apparent at the time of writing: the reluctance of people to self-isolate or restrict their social lives, in spite of the risks, to the point of police and military enforcement being necessary. This is what the authorities are up against: our social nature is a million-year-old evolved adaptation, not only driving individual behaviour but also underpinning the structure of society. Such is the social power of the brain to drive us together – and such is the addictive reward of social experience.

Beyond loneliness

Nearly fifty years ago, Robert Weiss described loneliness as 'a gnawing chronic disease without redeeming features'.[7] But we shouldn't be too harsh, or lay too much blame at its door. Loneliness is part of us, because we are inherently social beings – and *that* redeeming quality is one of the big secrets that have propelled us on the evolutionary journey to our status as the most successful species on the planet. Evolution did not invest too much in getting us beyond reproductive age. We are now faced with far, far longer lives than our forebears, and this entails the potential to be plagued by loneliness for many years – which is not good for us. It's certainly not good for our social brains, to a startling degree. But the good news is that loneliness no longer has to be endured; we can escape from it, and in doing so enjoy the prospect of better brain health.

Sex on the brain

ORGASM. BRAIN LIGHTING UP like Vesuvius. Pleasure radiating through-out the body – gripping, pulsating, addictive. From the hypothalamus, deep in the brain, bursts of powerful hormones flood into the circulation, generating a glow of well-being which lingers for minutes, hours, days. If the world seeks pleasure, it is the brain that provides it. Sex is all in the brain.

Of all the appetitive drives and behaviours – hunger, thirst, sleep – sex is arguably the most powerful. Driven by the deepest parts of the primitive brain, its impulses are mitigated only by our frontal lobes – via the learned social values which reside there and suppress an otherwise rampant limbic system. The limbic system, as we saw in chapter 1, tells you to do everything your mother told you not to – RIGHT NOW. It is the seat of the emotions: rage, aggression, fear – and sexual demand. To this extent, we might agree with Freud, who with more clarity than scientific justification, positioned us as 'simply actors in the drama of [our] own minds, pushed by desire, pulled by coincidence'.[1]

Sex has achieved this lofty (some would say torrid) status via the single most uncompromising principle of evolutionary biology: the imperative of transmitting our DNA. Everything and anything in the world of living creatures is predicated on this single, overarching principle: reproduce or die. And everything about human sexual behaviour, moulded by the mores of society over thousands of years, is driven by this impulse.

Such is the power of the human sex drive – a drive recognized indisputably by all cultures – that one feature common to all societies, past and present, is the presence of moral codes and social structures by which sexual behaviour is regulated. As we shall see in the following pages, these rules are dictated by how each culture understands the human condition and how it views society. This includes its view of sex and health – including brain health – and the differing nature and gender roles of men and women, including their sexual responses.

Sex-crazed women

In the great tragedies of ancient Greece, played out in the open before vast audiences of up to twenty thousand male and female citizens, important social messages were conveyed through sober acts of drama, interspersed with hilarious interludes of comedy. Satyrs would appear, half-man, half-beast, shamelessly boasting their huge erect phalluses as they regaled the audience with mockery of the play's principal characters, turning the mood and lifting the audience into fits of foot-stamping laughter. This comedy was a perfection of parody, because it was itself laden with social mores and subtle, unwritten messages: messages that conveyed meaning – and, with meaning, mechanisms of the social control of sexual roles and behaviour. It was the small penis, as in the great statues of Jupiter and Zeus, which conveyed the primary values underlying the Greek ideal of male beauty. Potency did not come from a large penis – which, as parodied in the comedies of Aristophanes and others, was seen by the Greeks as a symbol of stupidity and the absence of restraint. No: potency came not from a large phallus but from the intelligence – the brain power – required to drive parenthood. And, in the vocabulary of Aristotle, sustaining the *oika* (household), the basic unit of the *polis* (city-state) was everything. Maleness was a distillation of a man's ability to dominate others.

The Greek woman of ancient times was demure, sheltered, virginal until marriage and subdued, living in high respectability, 'segregated from the undesirable company of strange men'. Or so twentieth-century scholarship would have us believe, including the feminist writers of the 1960s, reinforcing a view of the subjugation of women. But if we consult another source, the medical texts of ancient Greece, we find a much less idealized view. The early Greek physicians, embedded in a culture which venerated physicality, beauty and innocence around the body, recognized the Greek woman as a highly sexual, though still respectable being – a view not entirely at variance with that of Aristophanes, whose female characters were vivacious and passionate, liked to socialize and to drink, and had a fondness for lewd jokes. Furthermore, for several hundred years from the time of Hippocrates in the fifth and fourth centuries BC, Greek physicians were united in the view that regular sex was necessary for a woman's good health, while continued virginity was harmful – a view entirely compatible with Athenian law, which went to unusual lengths to ensure that no woman was left without a man. In his seminal paper 'Aristophanes, Hippocrates and sex-crazed women', the classicist Konstantinos Kapparis emphatically states:

Medical literature, comedy and the laws of Athens agree on this point, that women needed regular sex in order to keep their good physical and mental health and balance. The women of Aristophanes are respectable, and yet sex-crazed; this was their nature, and a man's penis was not just an instrument of pleasure, although it was unashamedly that too, but also a necessary accessory for a healthy life.[2]

This benevolent view of the influence of sexual activity on health extended to 'mental balance', itself achieved by a proper balance of the four humours of the body (blood, phlegm, yellow bile and black bile), as discussed in chapter 1. At the time, it was believed that the loss and exchange of bodily fluids during sex helped the body's humours to maintain their equilibrium – which was, in turn, the basis of both physical and mental health.

Greek vase – the art of seduction

In summary, even at a time when physicians and the society they served knew nothing about the brain, they recognized the compelling nature of the human sex drive, the need for sex in both men and women, its relationship to health and well-being, and its key role in society. And they also recognized that good mental health depended on balanced sexual activity.

Death by celibacy

It's arguable that in today's society we have come full circle in our views on sexual behaviour and health to return to the ancient Greek espousal of tolerance and balance. Even so, while the sexually repressive views of the Church evident in the prohibition on 'lust of any kind', clerical celibacy and the derogation of women are undeniable, the idea that a balanced sex life offered benefits to health was surprisingly common in medieval times. So strong, indeed, was this view that the poet Ambroise wrote of death by celibacy at the Siege of Acre in 1189:

> . . . in pilgrims' hearing I declare
> A hundred thousand men die there
> Because from women they abstained.[3]

Such views applied equally to women, who it was believed should shed their seed through regular sexual intercourse to remain healthy. A woman who was not sexually active would succumb to suffocation of the womb, fainting and shortness of breath. Even more surprisingly for the time, some physicians recommended masturbation as an alternative to intercourse, testimony to the persistence of the Greeks' view of its value to good health.

The identification of moral virtue with sexual restraint persisted, nevertheless, through post-Renaissance times and into the Victorian era – a time of restraint in which women were warned against enjoying sex because of the alleged buildup of harmful 'nervous force' sexual pleasure would entail. As one nineteenth-century medic put it: 'Any unnatural performance of this act' – by which was meant anything other than the 'missionary position' – 'is apt to impair the health of the female.'[4] The enduring power of this sanctimonious and some would say hypocritical Victorian legacy is revealed by the poet Philip Larkin's emblematic statement: 'Sexual intercourse began in 1963 . . . between the end of the "Chatterley" ban and the Beatles' first LP'.[5] So it was that the modern sexual revolution began, and with it an understanding that our sexual lives should be guided by science and medicine, as well as morality. It was the pioneering work of two research centres in the USA which propelled us into our modern era.

From Kinsey to Masters and Johnson

Over a period of some fifteen years in the 1930s and 1940s, Alfred Kinsey and his team, based in the Department of Zoology at Indiana University, interviewed eighteen thousand people in the first ever investigation into the sexual habits and responses of men and women. Publishing their findings in the late 1940s and early 1950s, they reported that, on the basis of their sample:

- 10 per cent of men were predominantly homosexual;

- 2–6 per cent of women aged between 20 and 35 were homosexual;

- 92 per cent of men and 62 per cent of women had masturbated;

- 69 per cent of men had had sex with a prostitute;

- two-thirds of men and half of women had engaged in pre-marital sex;

- 50 per cent of women had orgasmed by age 20, 90 per cent by age 35.

So what, you may ask?

At the time, the publication of Kinsey's two reports was met with public shock and outrage, with indignation and disgust not far behind. Such things were not spoken of in decent society, much less made the subject of so-called scientific study. In the USA, publication was even thought to be a threat to national security. If you find this risible, consider what author Cami Beekman has to say: 'In the Cold War era, human sexual desire of any deviation from the marital norm would have been considered incorrect and dangerous to the strengthening of the American people for the protection of the free world.'⁶ If only mid-century Americans had known, like the ancient Greeks, the importance of sex to physical and mental health. In the event, these unprecedented reports not only revealed a reasonably accurate picture of contemporary sexual activity, but led to both a much greater openness about sex in public discourse and the beginnings of advances in the rights of non-heterosexual people.

Other researchers went beyond merely asking questions. In 1957, at Washington University in St Louis, William Masters (a gynaecologist) and Virginia Johnson (a psychologist and Masters' research assistant) had the audacity to investigate the physiological responses of men and women during sexual activity itself. Unsurprisingly, their research provoked huge controversy – not only because of the restrictive social conventions at the time but because, at Masters' request, Virginia Johnson had intercourse with him, so they could serve as the subjects of their own study – and eventually they became lovers. This arrangement is not what one might call conventional scientific method. More controversy followed their selection of other subjects: they began their experiments with 145 prostitutes, recruiting others – 382 women and 312 men – only later. But for all the furore their research aroused, their basic findings were groundbreaking and remain valid today.

Measuring responses during both masturbation and sexual intercourse, Masters and Johnson used novel physiological and other contentious methods which one could not imagine being used today: 'Men and women were designated as "assigned partners" and arbitrarily paired with each other to create "assigned couples".'[7] They used pioneering methods of measuring heart rate, blood pressure and breathing – and devised 'Ulysses', a transparent and self-illuminating dildo with a magnifying glass at the end, to measure the internal responses of their female participants. Their results enabled them to describe the 'four-phase human sexual response':

- excitement (initial arousal),

- plateau (full arousal),

- orgasm (climax) and

- resolution (post-climax recovery).

This cycle might now be described as 'old news' but it is still the basis of current studies into what is perhaps the last taboo in sexual research.

Today, investigators in the USA are conducting large-scale research into the secrets of female pleasure, the evolutionary driver of female behaviour and, as we will see, a response system deeply embedded in the brain. Their studies at Indiana University have so far involved over twenty thousand women aged between 18 and 95. An early study of 1,050 women asked them what makes a good lover. The three most common features were:

1   He/she takes the time to find out what I like (91 per cent of women).

2   He/she is attentive to me – listening and being aware of whether I'm enjoying sex (89 per cent of women).

3   He/she asks me what feels best for me (81 per cent of women).

In another study of two thousand women, researchers created a database of a dozen or so techniques that have led women to orgasm, including familiar terms such as 'edging', 'orbiting' and 'repetitive motion', as well as terms we hear less frequently (like 'staging', 'signalling' and 'framing'). Which techniques work best, according to the research, is highly personal. Readers may find details on the website 'OMGyes',[8] described by the *Sunday Times* as 'Nothing less than the next wave of an unfinished sexual revolution'.[9]

How much sex are we having?

Some younger readers may be surprised to learn that sexual appetite does not disappear at 40 – or even at 50, 60, 70 or beyond. The statistics speak for themselves, even suggesting the opposite: younger people are more dissatisfied with sex and are having less of it. In 2012, in the most recent of three surveys since 1991, the UK National Survey of Sexual Activity and Lifestyles found that less than half of men and women aged 16–44 have sex at least once a week, and the data suggested that nearly one-third had not had sex in the previous month. Moreover, in a study in the USA in 2010, middle-aged women (aged 31–45) were found to be more sexually active and to enjoy better orgasms than younger women (18–30). Not to be outdone, Public Health England tells us that English women in the 55–64 age bracket are the most contented with their sex lives compared to other age groups. Indeed, one intriguing study claimed that women over 40 actively seek sex more than their younger counterparts.

Generally, studies have shown that although the frequency of sexual activity does decline with age, it is not age itself that is the principal factor. Overall health and relationship status are more closely related to the frequency of sexual intercourse. In 2018 Dr David Lee at Manchester University published data from over seven thousand questionnaires in one of our most important surveys – the English Longitudinal Study on Ageing. His was the first national study of sexual health to include adults over 80, and it debunked many myths concerning the importance of sexual activity as we age. More than half (54 per cent) of men and almost a third (31 per cent) of women over the age of 70 reported they were still sexually active, with a third of these men and women having frequent sex – at least twice a month. Interestingly, sexually active women over 70 reported decreasing levels of dissatisfaction.

How may we explain these findings? There are some theories. One proposes that the combination of pressures on the 'sandwich' generation of young middle-aged people – work, caring for parents and bringing up children – are such that their sex life suffers. Others have related the decline in sexual activity among young people to the rise of the smart-phone and social media, virtual sex replacing real sex. This particular trend is found in all leading economies, including Japan, Finland, the USA, Australia and Britain. But it is not just about the percentages. What is worrying is the potential decline in human connectedness they suggest. All studies have shown that it is not so much the frequency but the mean-ingfulness of the sex that matters. Your life is not doomed if you don't have sex so often. Love and belonging matter more – with one caveat: as we will see, regular sex – at least once a week – appears to be the threshold for benefits to health, including the health of our brains.

Sex, health and longevity

Not all cultures are as far-sighted and sagacious as the ancient Greeks. The French call the orgasm *le petit mort* (the little death), reflecting the long-held view in Europe that sex diminishes the body – a view which conveniently invigorated the social strictures of the Church: 'It is good for a man not to touch a woman.'[10] In England, many male readers of a certain age will recall the now laughable myth that 'masturbation sends you blind'. In traditional Indian and Chinese cultures, ejaculation was viewed as a drain on vitality. As men grew older, they were advised to ejaculate less and less. So we must thank science for giving us indisputable, myth-busting evidence over the past fifty years showing that regular sex (including masturbation) throughout life contributes to good health and longevity. And the converse also holds true: throughout life, good health maintains sustainable sexual relationships. Let us take a look at that evidence.

Most men will breathe a sigh of relief to know that, though death does happen during sex, orgasm is very rarely the cause. A German review of 21,000 autopsies in 2006 revealed orgasm as the cause of death in only thirty-nine cases, and almost all of those involved sex with a prostitute (you have been warned!). By contrast, the results of the next study will comfort, if not encourage, all men (if they need any encouragement). In the forty-year Caerphilly Longitudinal Study, which started in 1979 and involved 918 men aged 45–59, the risk of mortality was found to be 50 per cent lower in those with a higher frequency of orgasm. It is what we call a 'dose-related' relationship or 'dose effect'. In other words, for men at least, the more orgasms you have, the lower your risk of death. With true British understatement and probably a little wry humour, the authors concluded that 'If these findings are replicated, there are implications for health promotion programmes.'[11]

There is more. Of the 918 men involved, sixty-seven had died ten years later from heart attacks and eighty-three from other causes. The researchers compared the men's reported frequency of sexual activity with their death or survival, and found that the death rate among those who had sex twice a week was only half that of the men who had sex once a month. It was another dose effect: as an individual's sexual frequency increased, so his risk of death decreased. Of course, the critics leaped on these findings: sex is related to good health, they said, so it wasn't the sex that extended longevity – it was the reverse, it was general good health that led to long life, with frequent sex a mere side-effect.

Unfortunately for the critics, there were no significant differences in smoking habits, weight, blood pressure or heart disease between the men with the highest and lowest frequencies of sexual activity. So, the most sexually active were not significantly healthier. The inevitable conclusion, therefore, was that, in middle-aged men, regular sex prevents mortality – another ticklish message for the health promotion industry. The most that the NHS can bring itself to say is: 'Weekly sex might help fend off illness.'[12] Importantly, it must be added, this doesn't just hold true for men. A Swedish study of 166 men and 226 women showed that in both groups the risk of death rose significantly if sexual activity ceased.

Why should it be that more frequent sexual activity extends longevity? One reason is that it has been shown to confer many metabolic advantages: it is associated with slimness, better heart-rate responses and lower blood pressure. It is often said that 'sex is as good for you as a three-mile run' – and this assertion has in fact been tested. Scientists in Canada studied twenty-one heterosexual couples aged between 20 and 30. They monitored their energy expenditure during sexual activity, and compared it to their energy expenditure when they ran for thirty minutes on a treadmill at a moderate pace. The results? The men burned 101 calories (4.2 calories per minute) on average during sex; women burned 69 calories (3.1 calories per minute). The average activity level, measured in METs (introduced in chapter 2), was moderate – 6.0 METs in men and 5.6 METs in women. By contrast, during running, men burned 9.2 calories per minute and women 7.1 calories. So, sex is not quite as demanding as running – but nevertheless does constitute reasonable aerobic exercise.

A recent interesting study in Denmark has shown a 'reverse effect': exercise improves sexual performance in terms of erectile function. These researchers found that forty minutes of moderate to high-intensity physical activity, four times a week for six months gave rise to an improved erection. If that were not enough, there is another self-reinforcing mechanism: regular sex has been shown to improve testosterone levels in men, which in turn improves sexual performance. Regular sex also improves the female reproductive cycle, for example, moderating menopausal symptoms in women.

There is another intriguing possible explanation for the longevity effect of frequent sex. One genetic study involving 129 women aged between 20 and 50 found that those who had regular sex had longer telomeres (telomeres are caps on the end of DNA molecules which prevent fraying – longer, non-frayed telomeres are related to longevity). But this is only one possible contributory factor; it's likely that the intimacy and well-being associated with partnered sex and the associated relaxation play a significant role. Many, many studies show this relationship.

The lovers

In his authoritative medical tome of 1898, *The Marriage Guide*, subtitled *A plain, practical treatise for popular use*, Dr F. Hollick firmly asserts one uncompromising message: sex for anything other than the purpose of procreation within marriage *is harmful to the human condition, even to the point of death*. Indeed, without daring to mention the term 'masturbation', Dr Hollick manages to dedicate a whole chapter to its evils. He summarizes the effects of this perfidious habit in both men and women as 'great lassitude and depression . . . most generally the memory soon begins to fail, and the mind cannot be directed to one thing for any length of time, but wanders continually; sometimes it even becomes unsettled altogether, and complete fatuity results'. Society outwardly endorsed these messages. I say 'outwardly', because it is widely estimated that in London at the time, among a population of 2 million, there were approximately fifty thousand prostitutes, undoubtedly offering this service.

Before we become all smug about 'twentieth-century enlightenment', it should be pointed out that though openness about sexual matters started with the Kinsey Reports, *it is only since the 1990s that science has been able to unravel the relationship between sex and mental well-being.* And what has this research revealed? All studies over the past twenty years, featuring participants of all ages, show there is a single incontrovertible finding: 'Sexual health, physical health, mental health, and overall well-being are all positively associated with sexual satisfaction, sexual self-esteem, and sexual pleasure.'[13] In short, feelings of mental well-being are higher among sexually active adults at any age.

Even the form of activity seems to make a difference, with studies showing that sexual intercourse is associated with better health and mental health benefits than other forms of sexual activity, particularly masturbation. Other studies go even further, showing that satisfactory sexual relationships and the benefits they confer, such as better mental health and less depression, are directly related to the *quality* of intimate relationships (closeness, passion and love) as much as to the *frequency* of sexual activity. These findings should not be surprising – research has revealed that orgasm is related to improved well-being and reduced stress, partly through the release of 'stress-busting' hormones such as dopamine and the 'love' hormone oxytocin. Also, the intimacy afforded by a sexually active partnership has been found to moderate levels of cortisol production – an important response to stress – and bring them within the normal range. After orgasm, a hormone called prolactin is released, leading to a feeling of relaxation and sleepiness. So don't blame your partner for falling asleep afterwards – it is perfectly natural!

Research by famed sexologist Beverly Whipple at Rutgers University has revealed that orgasm in women is a natural pain reliever, increasing the pain tolerance threshold by up to 75 per cent and the pain detection threshold by over 100 per cent. Some of these effects can be explained by the release of oxytocin and endorphins – natural pain relievers originating in the brain. Follow-up research using brain imaging is revealing more. Against all the odds, the researchers were able to recruit ten women who were prepared to lie in the noisy, cold, mechanical environment of an MRI tunnel and masturbate themselves to orgasm while being monitored. The resulting brain images showed an area of the brain 'lighting up' during orgasm – the dorsal raphe nucleus, known to release serotonin, to which the researchers attributed the pain-relieving properties of their orgasm.

Without wishing to diminish important current health messages, it's worth pointing out that another paper drew a surprising conclusion that is a smack in the eye for the high priests of health and safety: sexual intercourse without a condom is undoubtedly beneficial to women's health. To be frank, I was quite astonished when I first read the evidence; but it's unequivocal. In his voluminous review paper of 2010, Dr Stuart Brody concludes quite impartially that more condom use leads to lower mood, more depression (and more suicide attempts), a reduction in female pelvic reflexes, poorer vaginal health and diminished sexual responses. There are a number of explanations for these startling findings, all of them physiological or psychosomatic. Condom use reduces vaginal oxygenation and blood flow, and prevents contact with seminal fluid which contains mood-enhancing prostaglandin molecules. But that is not all. It also reduces intimacy and, with it, its stress-reducing effects on the brain. Brody concludes: 'Although there might be a direct chemical anti-depressant effect of semen absorbed from the vagina, the large difference in mood and suicidality might also be a result of intercourse with condoms not really being intercourse, but something akin to mutual masturbation with the same latex device.'[14] Many studies have shown the association of depressive symptoms with poor sexual responses.

Depression is the most common of all mental disorders, affecting an estimated 300 million people worldwide; it is also the most commonly diagnosed mental disorder of later life. In 2017/18, the recorded prevalence of depression in England was 9.9 per cent: that's nearly 5 million depressed people. It affects women more than men. In both, acute depression has negative effects on personal relationships as well as one's own well-being. Where the depression is organic – that is, where it originates in physical chemistry rather than in psychological factors – feelings of sexual desire often diminish because of neurotransmitter changes in the brain. In men, anxiety and low self-esteem, both symptoms of depression, can lead to erectile dysfunction – itself a driver of further depression. Erectile dysfunction may begin in men as young as 35 and affects 4.3 million men in the UK, about half of whom have sought no help. In every presenting case physicians should, according to the guidelines set out by the British Society of Sexual Medicine, take a full set of blood tests, including endocrine readings, and not prescribe any treatment for erectile dysfunction before a full differential diagnosis has been made, in order to rule out other conditions as far as possible. The good news for men (and their partners) is that, almost without exception, the condition can be resolved. No one needs to suffer in silence any more. What we see here is a classic case of a reverse paradigm: depression can cause sexual dysfunction, but even in cases not originating in depression, undiagnosed or untreated sexual problems can themselves lead to depression, anxiety, social withdrawal and other mental health problems.

Finally, we should look at a milestone study carried out in 2013 by a psychiatrist, Vicki Wang, and four of her colleagues at the University of California, San Diego. They recruited a sample of over six hundred older people, aged 50–99, all living at home with their partners. The research task was to understand if there was any association between sexual health and physical, emotional and cognitive function. Some surprising findings emerged. Over 70 per cent of these people engaged in sexual activity at least once a week, and over 60 per cent of them were satisfied or very satisfied with the sex they were getting. The study found that, in both men and women, symptoms of depression were negatively associated with sexual health, even after accounting for age, physical functioning, anxiety, cognitive ability and perceived stress. Further, it was concluded that depression was associated with worse sexual health more than with physical function, anxiety or stress, or age itself.

It's clear, then, that sex is good for your mental condition. What about chronic physical health conditions? We have already noted the metabolic advantages of regular sexual activity, including lowered heart rate and blood pressure; these, together with the reduction in stress levels associated with sex, are all conducive to protecting cardiovascular health. However, the picture is not that simple. Reporting on national longitudinal data gathered from more than two thousand American adults over 50, researchers at Michigan State and Chicago universities summed up their results as follows:

> We find that older men are more likely to report being sexually active, report having sex more often and more enjoyably than are older women. Results . . . suggest that high frequency of sex is positively related to later risk of cardiovascular events *for men but not women*, whereas good sexual quality seems to protect women but not men from cardiovascular risk in later life.[15]

This reinforces the message stated above: it is *both* the quantity *and* the quality of sexual relationships and activity that counts.

There is also a more specific point to be made here about prostate cancer, a condition not easily treated that kills over thirty men every day in the UK alone. In 1992, in what became known as the 'Harvard Ejaculation Study', thirty thousand men aged between 46 and 81 were asked the question: 'On average, how many ejaculations did you have per month during these ages: 20–29; 40–49; past year?' In the course of their responses, they provided details of their total ejaculations each month (from masturbation, intercourse and nocturnal emissions) for 1992. They were then followed to see how many of them were diagnosed with prostate cancer, with treatment and outcomes examined every two years up to the conclusion of the study in 2000. A surprising finding emerged: ejaculation frequency was linked to lower risk of prostate cancer. The incidence of prostate cancer was 31 per cent lower among men who ejaculated twenty-one or more times per month. For the non-mathematicians among you, that's basically sex of some kind every day with a couple of days off each week. A smaller but no less important Australian study ('Ejaculation Down Under') showed a similar result, with the strongest effect for the frequency of ejaculations in young adulthood, even though in these subjects prostate cancer was not diagnosed until decades later. These findings lack an unequivocal explanation, but are very encouraging in the effort to prevent a condition whose cause is largely unknown.

Box 7.1 summarizes the benefits of sex for physical and mental health and well-being.

BOX 7.1: THE BENEFITS OF REGULAR SEXUAL ACTIVITY FOR HEALTH, WELL-BEING AND LONGER LIFE

Chronic long-term illness
- For men, frequent ejaculation may protect against prostate cancer.
- For both men and women, regular sex protects against cardiovascular disease by lowering resting heart rate and blood pressure.

- Frequency of sexual intercourse has many metabolic advantages – it is associated with slimness, better heart rate responses, improved testosterone levels in men and moderated menopausal symptoms in women.

Well-being
- Those who have an active sex life at any age are happier and fitter.
- Sexual intercourse is associated with better health benefits than other forms of sexual activity, particularly masturbation.
- Orgasm improves well-being and reduces stress through the release of 'stress-busting' hormones such as dopamine and oxytocin.
- Orgasm in women is a natural pain reliever, increasing the pain tolerance threshold by up to 75 per cent and the pain detection threshold by over 100 per cent.
- The level of intimacy (closeness, passion and love) is directly related to the quality of sexual relationships.
- Frequency of sexual intercourse has been linked to better mental health and lower levels of depression.
- Depression, anxiety and stress have negative effects on sexual well-being.
- Conversely, undiagnosed or untreated sexual problems can lead to depression, anxiety, social withdrawal and other mental health problems.
- Sexual intercourse without condoms is related to improved mood and emotional well-being, and improved immune function, particularly in women.

Longevity
- Greater frequency of sexual intercourse is related to greater life expectancy.

The amazing sexual brain

The Greeks made much of the size of the penis, but does size really matter? Well, there is an intriguing tale to tell here, and it centres not on the Greeks but on twentieth-century science. And art. And above all on the brain. We are talking about a curious image, grotesque in its proportions, its depiction in both art and sculpture capable of giving offence, but critical to an understanding of sex and the brain. It is the homunculus.

The homunculus has interesting origins. In 1928, political infighting and intemperate feuding in the New York medical community – over a Rockefeller grant – led to an unintended but fortunate outcome. As a result of the unseemly mayhem, a little-known but rising star in neurosurgery moved out to Montreal – taking with him the millions of dollars of Rockefeller funding. So it was that Dr Wilder Graves Penfield, born in Spokane in 1891, developed a career based on the surgical correction of epilepsy. But this was to lead him into investigating hallucinations, illusions and déjà vu – even into the search for the human soul. It also led to one of the biggest breakthroughs in brain science: how the cells of the brain are organized in relation to the cells of the rest of the body. By stimulating the brain with minute electric shocks, Penfield worked out how the cortex was connected to the various other areas of the body – to groups of muscles working to produce complex movements, and to sensory nerve endings conveying our feelings of touch, pain, cold, heat and so on. From these observations he was able to construct 'motor' and 'sensory' maps of the brain, still in use today and largely unchanged nearly a century on. The most striking thing about these maps is that they portray a distorted image of the body – not as we see it in reality, but as the brain perceives it and organizes it into what is more and what is less important. Some body areas have lots of cells dedicated to them in the cortex; others have very few. Convert these brain maps into a three-dimensional image and you have the figure that Penfield called the homunculus, illustrated on page 202. In this depiction, the parts of the body are drawn in direct proportion to the numbers of their dedicated brain cells. The lips and mouth are large, showing that the brain dedicates very many cells to them. By contrast, the wrists and arms are skinny, showing that they have relatively few dedicated brain cells. In this way, the proportions of the homunculus body reflect the number of cells in the brain which deal with each area. We have very, very important hands. And genitals.

To summarize how the system works: the sensory cortex, positioned like the band of a headphone set near the frontal lobes, receives all types of information coming into the brain – visual, auditory, olfactory (smell), pain, touch, temperature, pressure and so on. Not all of what it receives becomes part of our conscious experience. Many of the responses of the body to incoming information are subconscious – what we call 'autonomic' (self-regulating) responses, such as those necessary to stabilize the body's internal environment (acidity, glucose, water, salt etc.). The overriding function of the sensory cortex is to coordinate responses, largely in the form of movement, via the motor cortex. And three of the sensory modalities it registers are vital for sexual activity – sight, touch and smell.

What does close inspection of the sensory cortex reveal about sexual function? An overgenerous investment in the sexual areas of the body – the genitals and the erogenous zones. What, we may ask, does this signify? In short, that there has been vast evolutionary pressure on the brain to ensure the survival of the species by maximizing sexual drive and behaviour. Pleasure is the inducement to mate. There are twenty thousand nerve endings in the male foreskin alone – and an equal number of dedicated sensory nerve cells in the brain. Each of those nerve endings picks up touch stimuli, and all twenty thousand of them send their messages about what's happening in the foreskin to twenty thousand nerve cells in the brain cortex.

We may wonder at the disproportionate largeness of the genitals in the male homunculus. But what of the female? Shockingly, while the first maps of the male sensory cortex were published in 1950, no meaningful corresponding research was carried out on the female sensory cortex for sixty-eight years after that. We could speculate on the reasons for that glaring omission. However, new brain imaging data now tell us that there is the same 'disproportionate' nervous investment in the cortical representation of the sexual organs and erogenous zones of women, with estimates of eight thousand nerve endings in just the tip of the clitoris. The clitoris is correctly described as the only part of the human body which serves pleasure alone. All women and most men will say that they know what the clitoris looks like. It may come as a surprise that most of them do not. All that is seen is the tiny external clitoris. The real clitoris is internal, a large sensory network extending to the labia, the vagina, the perineum and the anus. It is 5–6 inches long: the same size as the penis. The clitoris has therefore rightly been called the 'pleasure iceberg'. And all of the sensitive nerve endings of the clitoris serve the pleasure centres of the brain. It is a macro-nerve network for the pleasure of women. This revelation undermines Freud's distinction between the 'vaginal' and the 'clitoral' female orgasm, and is supported by a finding made some seventy years ago by that pioneering research duo Masters and Johnson: howsoever achieved, the physiology of the female orgasm is the same.

So here, depicted for the first time as an artist's illustration on page 202, is a female 'homunculus' – a 'femunculus' – to balance our perception of the sexual lives of men and women and to reveal the powerful underlying brain architecture that both sexes share.

In 2003, in the laboratories of Rutgers University, three women – all of them in wheelchairs – were reduced to tears. All had received spinal cord injuries; all had been told by their physicians that their sex lives were over. But in the course of an experiment to identify nerve pathways from the genitals to the brain, they orgasmed. For the first time in years. Imaging showed an alternative sensory pathway through the vagus nerve, which had never before been shown to extend into the pelvis in humans. It turned out to be the world's first evidence of where orgasms occur in women's brains.

Now imaging has shown us the exact location of orgasm in the female brain, and the sensory areas of the cortex devoted to the clitoris, vagina

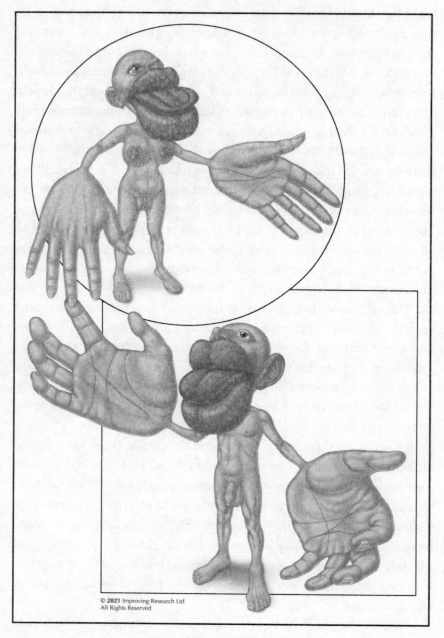

Femunculus and homunculus

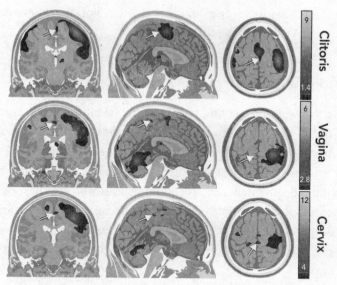

Figure 7.1: Sensory locations in the female brain (views from the front, side and above)

and cervix. These areas are shown in figure 7.1.

Multiple real-time studies have shown the extent to which orgasm activates the brain. It involves more than thirty integrated areas, all of which 'light up' during sexual activity, as shown in figure 7.2.

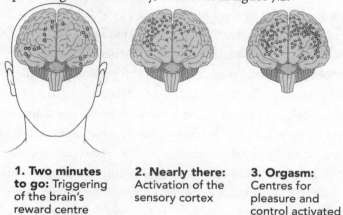

1. Two minutes to go: Triggering of the brain's reward centre

2. Nearly there: Activation of the sensory cortex

3. Orgasm: Centres for pleasure and control activated

Figure 7.2: Sexual activation in the brain

But not all orgasms are the same. For good evolutionary reasons, some are better than others.

Sex and brain health

What we know about sex and the brain, in scientific terms, is in its infancy. But it is dynamite. And it has implications for conventional public health messaging, preoccupied as it is with tempering sexual activity because of concerns about promiscuity, sexually transmitted diseases and unwanted pregnancies. Throughout history, sex has defied all worthy-minded attempts to suppress it. Because our need for sex is buried deep in the brain. It is a primordial drive cemented into place by millions of years of evolution. Sex is good for our health. And it is good for the brain.

Our story here starts not with humans but with the humble rat. Scientists looking at these small mammals made the intriguing finding that although exercise increases the production of the stress hormones known as glucocorticoids, it also has long-term health benefits and improves learning and memory. Such is its hedonistic value that a rat will not only 'ask' for access to a running wheel but will run up to 10 miles in a single night. It seems that the brain is protected against any potential adverse consequences of the exercise by a 'rewarding stress' – that is, one that requires a response from the body that lies within its available resources. We will deal with this idea in more detail in chapter 10.

Could the same be true of sexual activity? It is known that the sudden introduction of an unfamiliar male rat to a receptive female is stressful for both, though it is at the same time ultimately rewarding for the male (it probably won't surprise female readers that no male refuses the opportunity), and the female cooperates because she is in a receptive state. But what did the sex encounter do to the brain? When male rats were allowed a single sexual experience with a receptive female, their stress hormones were high. But in spite of this, there was increased growth of cells in the brain (in the cingulate gyrus, part of the limbic cortex involved with emotional control, learning and memory). A stressful but rewarding event had produced a benefit. But what about those rats who were allowed continuous day-by-day access to a receptive female? Their stress levels went down and the benefits to the brain increased, with more 'neurogenesis' – new cell growth – and more nerve connections. And the story gets even better. When scientists looked at older rats, they found that daily sex over a period of fourteen to twenty-eight days elevated neurogenesis to levels seen in the young adult rats in the 'control' part of the experiment. Sex was rejuvenating the brain. Other scientists have produced explanations for these apparent effects of sexual activity – for example, it stimulates the release of hormones, one of which, oxytocin, is known to activate cells of the limbic system (in the hippocampus).

Before we all start to get excited, we must ask: is there any actual evidence of these same side-effects in people? Early studies have shown that people with no age-related cognitive impairment are more likely to be sexually active and physically close to a partner than those with such impairment. In other words, it seems that if we look after and nurture our brains, we have a higher chance of sustaining regular sexual activity and intimacy. By contrast, research has shown that reductions in cognitive function (any kind of deficit in thinking skills or mental capacity) appear to be a barrier to maintaining or instigating intimate or sexual relationships. It is thought to be a matter of motivation – where brain function declines with age, those adults lose their desire for sexual activity.

But what about the reverse? Would frequency of sexual activity be of benefit to the brain? Little was known about this until 2016, when an Australian scientist, Dr Mark Allen, conducted a two-year study analysing data collected from six thousand adults over the age of 50 (average age 66). He examined this reverse relationship – how sexual activity might benefit our brains as we get older. He even offered up the possibility that sexual activity, as a rewarding social interaction, might offer some protection against cognitive decline (another challenge for public health messaging). His study looked at memory, sexual activity and emotional closeness. Why emotional closeness, you may ask? His reasoning was that previous human studies had identified attachment to and trust in a partner as important stimulants of hormonal responses such as the release of oxytocin – and oxytocin is known to promote neurogenesis in animal studies. Thus, the benefits of sexual activity for brain health might depend at least in part on the emotional closeness of sexual partners.

Allan's findings were remarkable. He found that those who were more sexually active *and* had greater emotional closeness had better memory performance. Moreover, there was a distinct age effect – the association between sexual activity and memory performance was stronger among the older participants, the 'milestone' age being 60.4 years. He also found that while memory performance declined for all participants over the two years analysed, this decline was unrelated to sexual activity or emotional closeness – it was purely an age-related effect. However, at any age, those who were sexually active had better memory function than those who were not.

Working along the same lines as Dr Allan, Hayley Wright and Rebecca Jenks analysed data from the English Longitudinal Study on Ageing; but they went further and looked not only at memory but also at mathematical performance, as a measure of higher executive function. They took data from an equally large sample, some 6,800 adults, aged 50–89 years. Across the whole sample, sexually active men and women were found to have significantly higher scores on both the maths and recall tests than sexually inactive men and women. Separating the sample by gender, however, yielded a surprising result – in women, those who were sexually active did not have a superior maths performance (number sequencing). The authors did not attempt to explain this gender difference, and further research would be needed to do so.

Later work by the same two authors, reported by *Science Daily* under the arresting title 'Frequent sexual activity can boost brain power in older adults',[16] widened the findings. In this smaller study, of the seventy-three participants – 28 men and 45 women aged between 50 and 83 years – thirty-seven said they had sex weekly, twenty-six monthly and ten never.

The participants' brain function, including attention, memory, fluency, language and visuospatial ability, was then assessed. The researchers found that those who had sex the most often scored on average two percentage points higher in visual tasks than those who had sex monthly and four points higher in verbal fluency tests than those who never had sex (each of these tests is a good measure of higher brain function). Curiously, there were no improvements in the other tests. As the authors acknowledge, we can only speculate as to the reasons behind these findings – and they offer their own speculations. Sex could be the cause of the improvements noted; it certainly looks as if increased sexual activity delivers a 'dose effect', but we do not know how or why. It could be that the secretion of 'well-being' hormones such as dopamine, oxytocin, serotonin and endorphins might be involved, and the authors suggest that this could be an area for further research.

But let's not confine our observations to those in middle age and over. What about younger age groups? Well, it should not surprise us to find that sexual activity appears to be beneficial to our brains at any age. In an intriguing study in Canada, research showed an association between the frequency of sexual intercourse in young women and improved memory. Scientists at McGill University in Quebec asked seventy-eight heterosexual women aged 18–29 years to report specifically on their frequency of sexual intercourse and to complete a computerized battery of mental tests which required them to distinguish previously presented faces and words from new faces and words. The results were enthralling. They showed that frequency of sexual intercourse was associated with the memory scores for abstract words *but not faces*. Our first observation should be that because the study was based on association only, it doesn't prove cause and effect: it is unclear whether better memory leads to more sex, sex improves memory, or there is another explanation for the association. But a second observation is really intriguing: memory for words depends largely on the hippocampus, whereas facial memory relies to a greater extent on other parts of the brain. Indirectly, this suggests that neurogenesis in the hippocampus might be higher in those female humans who have sexual intercourse more often, in line with previous animal research.

I have already mentioned in passing some important hormones which mediate sexual responses in the brain, and now we will turn to look at two of them – testosterone and oestrogen – each of which plays a role in defining sexuality, and both of which are found in both men and women.

Sex hormones and the brain

Your football team is playing a bitter rival at home, in a typical 'needle' match. The home crowd vastly outnumbers the pitiful away support. Excitement and noise vibrate around the stadium. An air of hostility, even intimidation, hangs over the game. After a titanic struggle, your team wins. Home advantage has paid off. But not how you might think. It was not the crowd. Or the familiar pitch. Or the atmosphere. It was the level of testosterone in your team's players which gave them the edge. An unlikely explanation, you may think? Except that scientific research has shown that, as a territorial response, much larger episodic secretions of testosterone can be expected in the home side.

Testosterone has a bad name. It is an anabolic steroid, building muscle and bone and power. It is said to create aggression, even violence and sexual lust. 'Testosterone-fuelled behaviour' has become a handy media catchphrase. This is matter for an interesting debate. Although it is not widely known, therapeutic dose studies show little or no evidence of changes in aggression or mood in men receiving supplemental testosterone. And another thing we don't hear is that *testosterone is vital for healthy brain development and function*. In order to take part in the chemistry of the brain cells, testosterone is converted to a molecule which binds or 'glues' itself to receptors – 'docking sites' on the brain cell membranes; or alternatively it can be converted to oestrogen by an enzyme in the brain called aromatase. Both aromatase and these receptors are found in key brain regions involved in memory and learning, including the amygdala and hippocampus, where testosterone has been found to increase the concentration of nerve growth factor.

In men, testosterone levels decline at about 1–2 per cent per year after middle age. Cognitive (mental) function also declines generally with age. Low testosterone levels have also been observed in patients with Alzheimer's disease and mild cognitive impairment, which raises the question of whether declining testosterone is linked to declining cognitive function. There is plenty of evidence to show that the combination of low testosterone and poor cognition is more than a coincidence of age. For example, studies in cell cultures and animals have shown that testosterone acts to protect brain function, and other studies have found that testosterone levels are associated with memory, spatial processing and 'executive function' in the brain. A systematic review of the science by Olivier Beauchet, a French psychiatrist now at McGill University in Montreal, showed that low levels of testosterone in healthy older men may be associated with poor performance on some mental tests. The results of randomized controlled trials have been mixed, but generally show that giving testosterone as a form of male hormone replacement therapy may have moderate positive effects on some aspects of brain performance (e.g. spatial ability) in older men. Spatial ability is the capacity to understand, reason about and remember the relationships between objects and space – not only understanding the outside world, but also processing information from outside and reasoning with it. It has been suggested that treatment with testosterone may protect the brain as we get older. But before middle-aged male readers of this book rush out to buy testosterone supplements, it is worth pointing out that they have to be prescribed by a GP in the UK, so getting hold of them is not an easy process (though it's not impossible). The situation in the USA is markedly different: here, where one in four men aged between 30 and 79 have symptoms of low testosterone, replacement prescriptions have tripled since 2001 and revenues from sales of testosterone drugs now stand at $3.8 billion per year. In short, American physicians will more readily prescribe replacement testosterone for those patients who request it, while the British NHS recommends against using so-called 'bio-identical' supplements – that is, supplements which are identical, at molecular level, to substances occurring naturally in the body. Notwithstanding the willingness of prescribing physicians, however, the US Food and Drug Administration notes that in 2018 its investigators uncovered thousands of unreported adverse incidents – by which they mean any untoward medical occurrence – involving bio-identical supplements.

For women, the picture regarding brain function and female sex hormones is similar. Oestrogen has multiple beneficial effects on brain function, promoting neurotransmitter function, glucose metabolism and the development of new synapses, and slowing brain ageing. Current evidence also suggests that although oestrogen alone has no role in the treatment of Alzheimer's disease once established, it may delay its onset. However, the therapeutic use of oestrogen in hormone replacement therapy (HRT) is mired in controversy.

The Women's Health Initiative (WHI) was a $625 million research study funded by the US National Institutes of Health – the largest health research body in the world – involving 160,000 women aged 50–79. Launched in 1991, it set out to look at the causes of morbidity and mortality in post-menopausal women, and in particular at heart disease, cancer and osteoporosis. Its core finding, that post-menopausal hormone replacement therapy (HRT) increased the risks for all of these diseases, had an immense and profound impact. However, these findings did not escape criticism – and indeed, from some quarters, scathing rebuke. In 2006, one critic wrote:

> In summary, the findings of the . . . study should have been: no
> significant risks were found for cardiovascular disease, invasive
> breast cancer, stroke and venous thromboembolism. Instead,
> the . . . authors . . . concluded that post-menopausal hormone
> treatment increased the risks for all of these diseases. When this
> was released to the news media it resulted in considerable confusion

among patients and doctors alike and caused an untold number of women to go without potentially beneficial hormone therapy. The final consequences of these incorrect conclusions are yet to be determined, but it is likely that an untold number of women will suffer from diseases which post-menopausal hormone treatment could have prevented.[17]

Five years after the WHI began in the USA, a study with similar aims was initiated in the UK. Three august bodies, Cancer Research UK, the NHS and the Medical Research Council, together funded a study of the effects of HRT on women's health, analysing data from more than 1 million women aged 50 and over – unsurprisingly called the Million Women Study. Its major finding, that women who use HRT are more likely to develop breast cancer than those who do not, fundamentally changed national prescribing policy and became medical dogma.

Now the rub. Like the Women's Health Initiative, and largely unknown to the general public, this study has also been criticized as being grossly flawed, not only in the invalid conclusions that were drawn from the data but also in the way in which they were communicated to a hapless public (and to a less hapless medical community) – by smug banner headlines in all media outlets. By contrast, the authors' retraction, after their own re-analysis, of one of their principal conclusions – that HRT increased the risk of breast cancer – attracted hardly any media coverage at all.

What has all this to do with brain health? I would make two points. First, the way in which evidence with potentially highly serious consequences was publicized leaves much to be desired and hardly shows the protagonists in a good light. It is a graphic reminder that many claims are made about what is good for the brain, and that not all of them are necessarily soundly based on unimpeachable evidence. As a scientist, I find this shocking. Second, progress in determining the usefulness or otherwise of HRT as an aid in preventing cognitive decline has not been helped by a prevailing medical orthodoxy and a more general atmosphere that discourage its prescription for both men and women. There is an urgent need for further research in this important area. HRT may well be good for women's brains – not just intellectual function, but emotional life and feelings of well-being. As a final note on this topic, it is probably worth mentioning a recent report on alternative 'natural' oestrogen supplements from the National Academies of Sciences, Engineering and Medicine in the USA, which stated: 'Marketing campaigns and celebrity influencers have touted claims of superior safety, effectiveness, and the "natural" and anti-aging properties of "bio-identical hormone therapies", but these claims of clinical utility are not substantiated by evidence from well-designed studies.'[18]

Now the $64,000 question – does HRT protect brain health? It's very complicated. Both oestrogen and testosterone are vital for normal brain development and for the maintenance of brain function, and both are implicated in brain ageing. Low testosterone levels have been observed in patients with Alzheimer's disease and mild cognitive impairment. Moreover, some studies show that the prevalence of Alzheimer's disease is significantly lower in women on HRT, and that women with Alzheimer's disease who were taking HRT had a milder form of the disease than those who were not. However, such evidence does not automatically translate into a recommendation to prescribe HRT (oestrogen and testosterone) for protection against neurodegenerative diseases: the evidence is as yet equivocal. And, given the current medical orthodoxy in Britain, few GPs are inclined to prescribe either. But it should be said that HRT in women remains the most effective solution for the relief of menopausal symptoms, including low libido and low mood, and is also effective for the prevention of osteoporosis. In certain age groups it may also provide protection against heart disease. In 2019, the renowned Mayo Clinic in the USA stated:

Women who begin hormone therapy (oestrogen plus progesterone) more than 10 or 20 years from the onset of menopause or at age 60 or older are at greater risk of the above conditions (heart disease, blood clots and breast cancer). But if hormone therapy is started before the age of 60 or within 10 years of menopause, the benefits appear to outweigh the risks.[19]

There are legitimate, safe and entirely legal ways to access these treatments in the UK, for example through bona fide medical consultants in specialist centres that have been audited successfully by the Care Quality Commission. Readers can access these on a private basis while remaining under the care of their local GP. One caveat is necessary: there are many convincing charlatans in the health marketplace, selling phoney, 'snake oil' hormone replacement therapies, especially for men. Great care should be taken to access such medication only through genuine medical specialists.

The key points of what we actually know about sexual activity and brain health are summarized in box 7.2.

BOX 7.2: WHAT WE KNOW ABOUT SEXUAL ACTIVITY AND BRAIN HEALTH

- Having no age-related cognitive (mental) impairment is associated with more sexual activity and physical closeness with a partner.

- Reductions in cognitive function (any kind of deficit in thinking skills or mental competence) are a barrier to maintaining or instigating intimate or sexual relationships.

- In animal models, frequency of sexual activity is associated with the growth of new brain cells, especially in the hippocampus.

- At any age, sexually active adults are more likely to have better memory.

- Greater personal bonding and emotional closeness during sex are associated with better memory.

- The enhancing effect of both intimacy and sexual activity on memory is greater as you get older (especially over 60).

- Sexual activity does not appear to stop age-related memory decline – but at any age, memory will be better if you are sexually active.

- Adults aged 50–89 who have regular sexual activity perform better on mental tests than those who do not (women on memory tests and men on both memory and number tests).

- Frequency of sexual intercourse is associated with better verbal skills.

- The principal sex steroid hormones, testosterone and oestrogen, have beneficial and protective effects on the brain throughout life.

Just do it . . .

Throughout this chapter, we have talked rather loosely about 'sexual activity' without really defining it. It can vary from masturbation through mutual stimulation to full sexual intercourse. The fact is that researchers define 'sex' almost arbitrarily with respect to their various studies, and there is little homogeneity across the research. There are a number of very genuine reasons for this inconsistency – culture, age distribution, health, research aims and so on. Even so, as a general conclusion, it is reasonable to say that while almost any kind of sexual activity has been shown to benefit brain function, a good rule of thumb for the assurance of benefit appears to be a frequency of 'once per week'. Certainly, many researchers appear to have shown a 'dose effect': that is, the more sex you have, the more benefit you are likely to derive from it. And this extends from general health to brain health. So, a message that may very well not be unwelcome: have sex regularly for optimal benefit to the brain.

During the course of my research for this book, it became clear that one form of sexual activity stood out as the vehicle of optimal benefit: full sexual intercourse within a close intimate partnership. Even as a physiologist, I found some of the discoveries I came across astonishing – for example, that the composition and quality of semen ejaculated during masturbation is inferior to that emitted during sexual intercourse; that not all orgasms are the same, most pleasure being generated during intercourse; that interference with the natural conditions of intercourse, for example the use of condoms, removes distinct physiological and psychological benefits, particularly for women; and that emotional closeness in particular drives distinct and material benefits for brain health. There appears to be almost a sexual hierarchy of benefit: don't masturbate if you can have intercourse; and if you can have intercourse, it is better in an intimate and close relationship. *Voilà!* It is as if our evolutionary forebears have provided us with a blueprint for the extraction of brain health from the sexual relationships most suited to maximizing human progress and survival.

It's commonly believed that sexual activity will decline as we leave our twenties and thirties behind us. There is no doubt some truth in this, but there are two caveats to be made. One is that, as shown earlier in this chapter, the decline in frequency of sexual activity with age is nowhere near what people generally believe – our active love lives can persist into older age, even until our seventies and eighties. The key to maintaining an active sexual life is twofold: be sexually proactive, which generates higher levels of sex hormones and maintains libido; and do everything possible to maintain every other aspect of health, through diet, exercise, remaining socially active, getting good sleep and avoiding excess alcohol consumption. The last point is particularly important for men. Neither our brains nor our testicles will 'forgive' high levels of binge drinking in our twenties and thirties, and even beyond. Saturating the body with alcohol on a regular basis destroys libido and eventually leads to impotence (we will look at the effects of alcohol on the brain in more detail in chapter 10). But the good news is that following these rules need not be onerous; and the evidence is that maintaining good general health and an active sex life are not only beneficial in themselves but are hugely beneficial for brain health. Box 7.3 summarizes the main recommendations.

**BOX 7.3: SEXUAL ACTIVITY AND BRAIN HEALTH –
RECOMMENDATIONS**

- Actively make the effort to be sexually intimate with a partner
  and be proactive together; set a target of at least once per
  week. This will have benefits for lifelong brain health.

- Take steps to maintain physical and mental health across
  adulthood – eat a healthy diet, don't be overweight, develop
  good sleep habits, don't smoke tobacco, don't drink too much
  alcohol, avoid harmful stress, take regular exercise and keep
  physically fit. Both physical and mental health are related to
  maintaining regular sexual activity.

- Do not sow the seeds of poor sexual performance by excessive
  drinking! Remember that excessive alcohol consumption from
  youth onwards has a lasting effect on both the brain and on
  sexual performance, particularly in men. Many tissues will
  'forgive' the alcoholic excesses of youth but the testes are not
  among them.

- Regular sexual activity will improve levels of testosterone in
  men and oestrogen in women, in turn improving the chances
  of regular sexual activity – a 'virtuous circle'.

- Maintaining regular sexual activity will improve feelings of
  well-being at any age. To help maintain sexual activity as you
  get older, consult a physician who can offer specialist help,
  especially for erectile dysfunction in men, and may prescribe
  HRT for both men and women.

- After the age of 60, make a special effort to have partnered
  sexual activity – with physical and emotional closeness. Sexual
  activity and emotional intimacy will confer benefits in memory,
  higher executive functioning and other mental skills.

- In men, higher levels of testosterone are associated with better
  cognitive function, including memory and higher executive
  function. Though promising results have been found from
  taking supplemental testosterone, it isn't possible to say with
  certainty that it will improve thinking skills or prevent cognitive
  decline, including dementia, though men taking HRT do report
  improvements in well-being.

- In women, the sex hormone oestrogen appears to have a protective effect against cognitive decline and dementia. However, after the menopause, oestrogen levels decline. As with men, it isn't possible to recommend HRT for overall improvement or maintenance of cognitive performance in older post-menopausal women, though women taking HRT do report improvements in well-being.

In conclusion to this chapter, it should have become clear to the reader that maintaining sexual activity throughout life is important at any age, for both general health and brain health; we are fortunate that so many of the research findings in this area have been established in the past few years. It is manifestly also a reciprocal arrangement: good brain health and good general health are conducive to sustained sexual activity; and in turn, sustained sexual activity has immense and only recently explained benefits for the brain. In 1988, one of the most commercially successful slogans ever was launched by the sports firm Nike. Directed at all consumers of any background or age, it was designed to be universal and intensely personal. I can think of no better directive to sum up the messages of this chapter – and it's one that would be immediately understood by Aristotle, Hippocrates and the respectable, sex-crazed women of ancient Greece: JUST DO IT.

Mind games

EVERY DAY IN THE brain, hundreds of new cells are produced in a process called neurogenesis. From animal studies we know that – regardless of how many are produced – about half of these new cells will die within one to two weeks. And they die before they have managed to connect to existing brain cells. This programmed cell death is not inevitable. It can be stopped. *By brain training.*

Can we 'hack the brain'?

In the complex, technologically driven world in which we live, heavy demands are placed every day on our thinking skills, on our memory and on our concentration. If we cannot meet these demands, we will fail to keep up with others and with their expectations. This is a fundamentally different environment from the one in which our brains evolved. Now we must process vast amounts of new kinds of data and learn the skills to cope with all these inputs. Some have called the acquisition of these new skills 'hacking the brain' – developing ways by which to overcome our natural limitations. These pressures are compounded by the ageing of Western societies. As we have discussed, losing our mental sharpness as we get older has become one of the biggest health-care and occupational concerns, as we experience continued day-to-day demand on our faculties.

It therefore should not surprise us that the idea of 'brain games' that can increase our brain power has seduced the public mind – and purse. In 2005, consumers in the USA spent $2 million on 'brain training' games. By 2013, that sum had risen to $300 million; and in 2014, it reached a staggering $1 billion. Sales volumes are similarly huge in Europe, the UK and Asia. Extravagant claims have been made about the benefits such games yield – to the point where some have given rise to court cases.

These games are not just marketed at adults – some are aimed at babies too. On a not-so-special evening in 1996, Julie Aigner-Clark, a 'stay-at-home mom' in Georgia, USA, had a special idea. Why not make a video for infants that stimulates their brains? Thus the *Baby Einstein* product was born. The first video was filmed in the basement of Ms Aigner-Clark's home and cost a mere $18,000. In 2001 she sold the rights to Disney for $25 million. She appeared on the *Oprah Winfrey Show*, *Good Morning America* and *USA Today*. She was lauded as a star entrepreneur by President George W. Bush in his 2007 State of the Union address. *Baby Einstein* was soon followed by the sequels, *Baby van Gogh*, *Baby Galileo* and *Baby Shakespeare*, each one claiming developmental benefit from the early exposure of the infant brain to the 'right' stimuli. By 2002, *Baby Einstein* was mesmerizing parents and infants across the USA. It is difficult to overstate its impact – it's estimated that one-third of all babies in the country at this time were exposed to one or more of these 'Baby Genius' videos. The series promoters proclaimed: 'It's true! Classical music and powerful images stimulate your child's brain.'

There was just one problem with this claim. There wasn't a shred of evidence to support it – not least because no research on the matter had been done. In 2007, a journal paper explained that for each hour spent watching baby DVDs/videos, infants understood on average *six to eight fewer words* than infants who did not watch them. Though this was contested by the founders, by 2009 the bubble had burst. Disney admitted that the videos had no educational value and sold the brand. There were no sequels.

The story of *Baby Einstein* is a salutary one. Commercial success (allegedly $400 million in total revenues) was derived from an idea which had little if any scientific merit but was intuitively attractive to parents. Such occurrences are not uncommon. Throughout this chapter we shall be examining the evidence for the merits of brain training, notwithstanding the intuitive attractiveness of the idea.

The tiny minds from which we have gained our clearest picture of the effects of training in the brain are those not of human infants but of animals examined in the course of scientific research on the newly generated brain cells mentioned at the very beginning of this chapter. Yes, programmed cell death in the brain can indeed be halted by 'mental training' – but not any old kind. The mental training has to have certain features. To keep the new cells alive, it must be *intensive*. It must require a *high concentration of effort*. It must lead to the *successful learning* of a new skill. And *it must be sustained* – through many trials every day, for many days.

If we look further at what works and what doesn't work, the story becomes even more fascinating. One type of very effective training involves what psychologists call 'associative learning'. This type of learning is common in humans. The principle behind it is that information and ideas that can be associated – linked together – are more easily learned. The brain was never designed to learn and remember things in isolation – it groups things together into one 'associated memory'. For example, we don't remember isolated features of someone's face – we remember *the whole face*. Interestingly, this principle was recognized over five hundred years ago by Leonardo da Vinci, who wrote, in *A Treatise on Painting*: 'All the parts . . . must be correspondent with the whole . . . I have experienced no small benefit when in the dark and in bed by retracing in my mind the outlines of those forms . . . by this method they will be confirmed and treasured up in the memory.'[1]

Associative learning is a form of conditioning – a term that describes how new behaviour is established by reinforcing it with a reward. A good example would be Ivan Pavlov's work with dogs to demonstrate their learning that a stimulus, such as the ringing of a bell, leads to a reward, or food. After a few iterations of this sequence of events, the dogs began to salivate as soon as they heard the bell, having learned that food would soon follow this signal. This kind of method was taken further some decades later by B. F. Skinner, an American psychologist of the twentieth century, who used it to shape the behaviour of animals – for example, he famously taught pigeons to play the piano by rewarding the appropriate key presses.

In animal studies, a second type of effective training involves 'spatial learning'. This is also common in humans. Imagine you are in a strange city. As you find your way around, you construct a 'map' in the brain, which involves your taking in and organizing information about your environment, to enable you to navigate successfully. You will remember all those locations which help you to find your destination – because these are rewarding. This is actually a complex, specialized form of associative learning. It is especially demanding because during it we have to create associations between many loosely related bits and pieces of information.

Researchers have also found through animal studies that learning *new physical skills* is very effective in consolidating the populations of new neurons in the brain. The more complex the physical task, the greater the effect. But there is a subtle difference between the effects of physical learning and of mental training. Physical activity, especially aerobic exercise, greatly increases the *production of new cells* in the brain. For example, in small mammals, researchers have found that two weeks of daily exercise result in an increase of about 50 per cent in the number of new cells produced in the hippocampus (centre of learning and memory). This effect has also been observed in human research, as we saw in chapter 2. By contrast, mental training increases *the numbers of neurons that survive*, rather than leading to the creation of new ones.

We know that once these three types of learning – associative learning, spatial learning and the learning of new physical skills – have taken place, the surviving neurons will remain in the brain for several months;

and by the end of that period, these new cells will have formed simple functional connections to other brain cells. But rescuing new cells from death is not that simple. There is one further condition. We know that there is a 'dose relationship' between how well an animal learns and the number of brain cells that survive. Learn poorly or not at all, and it's 'game over'. Likewise, if the training isn't very challenging, the new cells don't survive. However, if the training and the learning process are difficult, but the animal still learns from them: 'Bingo!' – the new cells survive. Those animals that learned the skill but required more training to do so tended to retain more new cells than animals that learned with less effort. The training, it seems, must be demanding if it's to be effective.

In summary (see box 8.1), animal models show that neurogenesis occurs in the brain, especially in the hippocampus, and that these new cells can be protected by certain types of brain training, as long as new learning takes place. Does this apply to us?

BOX 8.1: TRAINING THE BRAIN – LESSONS FROM STUDYING ANIMALS

Producing new brain cells and protecting them depends on:

- Type of training:
  - physical training: learning new physical skills stimulates neurogenesis;
  - mental training: associative and spatial learning protects the new cells.
- Intensity: the training must be intense, requiring multiple attempts to learn.
- Effort: high levels of concentration are needed.
- Practice: repeated and sustained levels of training are required.

- Success (reward): the more successfully the animal performs in the training, the more cells are preserved in the brain.

- Demand: the training must 'challenge' the brain so that new learning takes place.

Can the human brain grow new cells – and keep them?

In the human brain, as in that of small mammals, the hippocampus is a vital centre of learning and memory. It is, of course, not the only part of the brain involved in these functions – other equally important areas are the pre-frontal cortex (frontal lobes), the amygdala and the cerebellum ('little brain' – a vast coordination centre at the back of the brain). But the hippocampus is crucial – without it, forming new memories is impossible.

The human brain is able to change physically in response to the way it is used. This was demonstrated with spectacular clarity in respect of the hippocampus in an experiment involving black-cab drivers in London. Not just anyone can drive a traditional black cab. In order to get a licence, you have to learn 'The Knowledge' – a working memory of the myriad of the 25,000 streets within a 6-mile radius of Charing Cross railway station. Not the grid system of New York. Not the circular *arrondissements* of Paris. This means learning your way around the murky maze of back-street London – a formidable task that can take three or four years, mostly done by riding around the capital's streets on a bicycle or moped. Nearly three-quarters of all those who start the course drop out. Once you have passed, your daily work is a fearsome challenge. Anyone getting into your cab can expect to be taken faultlessly to any other point in London, without thinking twice about getting lost. And there's no satnav! In 2011, scientists at University College London decided to look at the 'black cab hippocampus'. Using MRI, they imaged the brains of black-cab drivers and a control group of non-taxi-drivers. They found that the cabbies had higher volumes of grey matter in the hippocampus than non-cabbies of similar age, education and intelligence. Going further, they tested seventy-nine apprentices over a four-year period while they were learning 'The Knowledge'. Of those seventy-nine, only thirty-nine graduated; and each of these thirty-nine not only had a bigger hippocampus than those who dropped out, they also performed better on memory tests.

The clever cabbie

One big reason for these differences is that new cells are being formed in the hippocampus in response to the huge and sustained challenge of learning 'The Knowledge'. In other words, neurogenesis is happening. And we can be sure that the long-term pressure of keeping up performance goes a long, long way towards consolidating that growth in grey matter.

There are wide individual differences in the size of the hippocampus. These differences are shown in figure 8.1 in the pattern of vertical dots showing the range of measurements for each age. This graph also shows two other things. First, as we age there is a trend for the hippocampus to shrink. However, this isn't necessarily a reason to worry. For the second point is, as you will see if you compare the data horizontally, looking at the range of dots for each age, that some people in their seventies and eighties are doing better in terms of hippocampus size than some in their forties and fifties. These observations reveal a cardinal principle of brain research, and a very welcome one at that: it is *not* inevitable that as we age, we experience a decline in brain power! And, contrary to the received wisdom that would attribute these differences to genetics, the evidence actually shows that lifestyle and environment are responsible for the greater part of them – in other words, how this part of our brain ages is largely within our control.

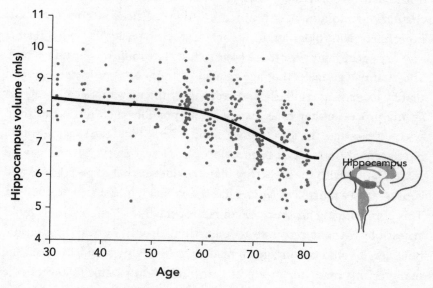

Figure 8.1: Hippocampus volume at different ages

And it's not just the hippocampus. A large and convincing body of research tells us that throughout our lives, the brain continues to develop new neurons and synapses in many different areas. These cells and connections are malleable, meaning that they are susceptible to change according to how we live our lives – so we can legitimately expect what we do to make an impact on how our brain adapts and functions. Actions we can take can affect how these nerve cells and synapses develop and how well our brain functions, from memory to attention, and from thinking to language and reasoning skills. One impressive piece of evidence has shown that neuroplasticity in the brain can be influenced by training and that the neuroplasticity becomes evident in the form of new brain cells. In a study carried out in 2009, Kirk Erickson and colleagues at the University of Pittsburgh showed that one year of regular aerobic physical exercise increased the hippocampus volume by 2 per cent – equivalent to reversing ageing by two years! – and also led to improved memory. The researchers showed that this change was brought about by the release of the neurotrophic factor BDNF – which, as we have seen in earlier chapters, stimulates new brain cell growth. Erickson conducted his experiment with older adults, people in their late sixties, so his results are very encouraging for those approaching late middle age and beyond. They tell us yet again that age-related changes are not inevitable, and that by exercising regularly and reasonably intensively we can keep the ageing process at bay to an extent – and possibly even reverse it. Even twenty years ago, the idea that we could reverse biological ageing would have seemed outrageous to most people. Now, however, it has gained such traction with the gathering of new evidence that so prestigious and conservative a journal as *Nature* has stated that it might well be possible. This is an enticing prospect, of course, especially as it increasingly looks accessible without recourse to medicinal drugs. By way of caution, however, I would encourage all readers to be highly sceptical about the many claims made for products that purport to be able to reverse the ageing process.

To summarize so far, then: we experience neurogenesis in our hippocampus, a vital centre of learning and memory, *and* in other important parts of the brain. These new cells and connections can be protected by our lifestyle, and the brain can become more neuroplastic. We can bring all this about by mental training and physical activity.

Cognitively stimulating activities

The next question must be: what kinds of training or learning experience help our brains – and how do 'brain games' contribute? The lessons that have been learned from research with small animals and humans alike point in one direction – whatever the training, it must be *cognitively stimulating.*

Cognitively stimulating activities are mentally engaging activities or exercises that challenge our ability to think. Examples are learning to play a musical instrument, taking up a second language, having dancing lessons, learning to play cards, and learning complex, new mental and physical skills, like tai chi or juggling. Juggling is a particularly interesting one: research has found that it is so complex and so demanding that it *alters the structure of the brain* – creating new tracts of white matter and increasing the volume of grey matter. It's as if the brain *wants* to be puzzled and learn something new!

These activities can help us to maintain and sharpen our cognitive abilities – brain functions such as decision-making, memory, thinking, attention and reasoning. Some have called such activities 'cognitive enhancement strategies'. But whatever they are called, they all share some essential features, as we saw in the animal research noted above: the activity must be challenging, it must be intense, it must require high levels of concentration and it must involve the successful learning of a new skill. In essence, it must be demanding if it is to change the dynamics of the brain. 'Cruising' is not an option.

Although not widely recognized as behavioural strategies, many techniques for enhancing brain function have been with us a long time, in some cases for centuries. They include well-established cultural activities such as musical training, learning to dance and taking up a second language. Some of these activities are thought to show benefits to our thinking skills beyond the specific skills being taught. For example, a meta-analysis of twenty studies done in 2014 showed that tai chi appears to improve executive function, as measured by mental testing – the ability to multi-task, manage time, make decisions – and to increase the volume of certain areas of the brain. In older adults with mild cognitive impairment, tai chi slowed down the rate of decline more than other types of exercise and improved cognitive performance. Let us now take a look at two other activities with high volumes of supporting research: learning a language and dancing.

As a system of communication, language is not only uniquely human but is inherently complex, diverse and exquisitely structured. It emerged at least 150,000 years ago, at a time of huge structural growth and massively increasing complexity in the brain as we became true humans, characterized by 'modern' human behaviour such as abstract thinking, planning, art, music, dance and group hunting. The scientific study of language's origins has been plagued with polemicism – to the point that in 1866, the Linguistic Society of Paris banned any debate on the subject, a prohibition which hindered meaningful investigation for over a hundred years. It remains one of the hardest areas of human anthropological research, and there has only been real progress since the early 1990s.

Language is so naturally embedded and widely integrated in the structures of the brain that it has been said that we cannot understand the brain unless we know how language contributes to brain function. It is integral to emotional expression, thinking, reasoning and remembering. Now new science is revealing there may indeed be benefits from being bilingual and learning a new language.

First, *learning a new language can increase the size of your brain*. In 2012, language students at the Swedish Armed Forces Academy were compared with other students not studying languages (the control group) while both groups undertook learning tasks of equivalent difficulty over a period of three months. Researchers imaged their brains at the beginning and end of the exercise. They discovered that the brains of the language students were larger in certain areas than those of the non-language students, which showed no increases in size. And what was the area most affected? You guessed it: the hippocampus! But the changes were not the same in every student – those who had found the tasks most challenging and had had to put in the most effort showed greater increases in size not just in the hippocampus but in other parts of the brain involved in learning, such as the middle gyrus of the frontal hemispheres. In summary, the biggest increases in size were dependent on how successful the students were and *how much effort they put in*. Just as in the animal experiments.

And what of those who have been bilingual all their lives? Researchers in Seattle compared a group of Americans who were bilingual in Spanish and English with a group of their monolingual compatriots, who knew only English, by imaging their brains to look at the differences in white matter (the communication fibres). In a complex report, they showed that 'foreign language immersion' induces neuroplasticity in the adult brain and that there was a 'dose effect': that is, the degree of alteration is proportional to the degree of immersion in the language. It appears that brain structures are different if you speak more than one language, and that the difference it makes to the structure of the brain depends on how much you use the second language. This finding is very important, because it explains in part why it is more difficult for us to learn a new language as adults than as children. Children's brains show more neuroplasticity than adults' brains. Learning a second language, however, improves the plasticity at any age. But there's another difference. Where the language is processed in the brain depends upon *the age at which you learned it*. Children under the age of 12 apparently default to a single storage area for both languages. Adults require a different area for each. It seems that the brain structures involved in learning a new language are not fixed but fluid, undergoing subtle changes as each new language is learned.

It's now also widely accepted that using a second language requires the brain to activate many different parts, including the frontal lobes, deeper sub-cortical areas and the corpus callosum, that wide band of tissue that allows the two halves of the brain to talk to each other. This is because both sides of the brain are involved, information being exchanged between the two halves. This increased activity increases the volume of our white matter *and* the number of fibres, improving the 'cross-talk' between the brain's two hemispheres.

And it's not just the white matter that changes as we use a second language. Scientists at Georgetown University in the USA compared the grey matter of the brain in bilinguals and with that in monolinguals. Their finding was unequivocal – bilinguals had more grey matter where it mattered, in the executive frontal lobes. It looks as if that 'cross-talk' – the effortless exchange of information between the hemispheres – generates a bigger volume of grey matter, the processing part of our brains.

Second, an even bigger question: does learning or speaking a second language boost our brain power? An intriguing experiment was carried out by two researchers, Dr Thomas Bak of Edinburgh University and Dr Suvarna Alladi of the Nizam Institute of Medical Sciences in Hyderabad. They compared two groups of students, one bilingual and one monolingual, and found that the bilingual students performed better on attention tests than the monolingual group and had superior concentration. Other work by Bak has also answered another fundamental question – do the benefits of a second language depend on the age at which we learn it? To answer this question he compared two groups of adults undertaking a Spanish language class – one group aged 18–30 and another whose members were all over 56. The results were surprising. Over the four-week course, everyone, regardless of age, improved on tests of attention, memory and fluid intelligence ('mental flexibility'). However, *the older adults improved more than the younger ones*. This finding is consistent with all previous research, which has shown that learning leads to cognitive improvement regardless of age, and that older adults improve more than younger ones over the same period of time. Remember the animal studies on mental training? It was also found that these performance improvements depended on how much individuals practised using the language.

There is yet another encouraging finding: learning a new language is also a great way to buffer your brain against ageing. For evidence of this, we need to look no further than Italy, where researchers scanned the brains of eighty-five adults who were all at a similar stage of dementia. Forty-five were bilingual (in German and Italian) and forty monolingual (in either German or Italian). The brain scans were used to detect glucose uptake, which showed activity in different parts of the brain – and how well these parts connected functionally to other brain regions. A remarkable finding emerged: on average, the bilinguals were five years older, but their disease was at exactly the same stage as the monolinguals – that is, it appeared that their linguistic prowess had slowed down the progression of the disease. The monolinguals' brains also showed slower metabolism in key brain areas, implying a greater levels of dysfunction, while the bilinguals had better connections between executive areas of the brain – and their immersion in their second language was significantly correlated to activity in key brain networks.

All this sounds good. However, a crucial question remains: do people improve their fluid intelligence by learning new languages, or are those with better fluid intelligence more likely to take up learning a language? Which way around is it? This question has been answered by the famous longitudinal study, The Disconnected Mind. As we saw in the introduction, the participants in this study all took an intelligence test in 1947 at age 11 and were retested in their seventies and early eighties. Two hundred and sixty of the participants spoke at least one language other than English. Of those, 195 had learned to speak it before age 18, and sixty-five after. The researchers found that those who spoke two or more languages had significantly better cognitive abilities than would be expected on the basis of their intelligence at age 11. The abilities that benefited the most were general intelligence and reading, irrespective of whether they had learned the language early or late.

The evidence, then, is conclusive: learning a second language, at any age, is truly a 'cognitively stimulating activity'. Not only does it 'rewire' the brain and increase its size, it is one of the most effective ways of increasing our brain power, keeping our minds sharp and 'buffering' our brains against ageing.

What if you prefer your cognitive stimulation to come with music and movement? If you want to stay young and increase your brain power, you might want to take part in Strictly Come Dancing. Dancers certainly think so. In 2017 at the University of California, Irvine, a survey was carried out of more than two hundred dancers (71 per cent of them female) from the Atomic Ballroom Dance Studios, to determine their perception of the physical, emotional, cognitive and social benefits of taking dancing lessons (if any). The results were salutary.

As far as cognitive benefits were concerned, a large majority – 82 per cent – agreed that dance improved their memory or their ability to learn new things, and 70 per cent that dance helped them to focus and pay attention for longer periods of time. An even bigger proportion (95 per cent) agreed that dance improved their emotional life. Social benefits were also rated highly, with majorities reporting that partnered dance made them more comfortable in making and keeping eye contact (80 per cent), in making physical contact (89 per cent) and in meeting new people (again 89 per cent). Participants reported less nervousness in social situations (another 89 per cent) and better social and interpersonal skills (88 per cent). And an overwhelming majority (93 per cent) agreed that dance helped improve their self-confidence.

There was an intriguing second group of findings, too: more experienced dancers reported significantly greater benefits in all the areas

Keep dancing

– thinking skills, emotional life and social life. Length of dance participation and frequency of dancing were two crucial determining factors.
Cognitive benefits were correlated with age – older participants indicating
greater cognitive benefits – but we don't know whether there was an
actual age-related benefit or whether older dancers were simply more able
to perceive the cognitive benefits.

Does research back up their perceptions? The answer is mostly
'yes'. Most research has shown that learning to dance is good for our
brains, especially in maintaining lifelong processing skills and speed of
processing. For example, in 2012 researchers at North Dakota's Minot
State University found that Zumba, a popular form of Latin-style dance,
is beneficial to mood and improves certain cognitive abilities, such as
visual recognition and decision-making. Other studies show that dance
reduces stress, increases levels of the feel-good hormone serotonin and
develops new connections in the brain, especially in regions involved in
executive function, long-term memory and spatial awareness (learning
to know where we are).

What lies behind these benefits? A new study, carried out in
collaboration by several universities in the USA, has found a possible
explanation. Over a six-month period they compared the effects on the
brain of several lifestyle activities, one of which was taking lessons in a
particularly demanding form of dance – country dancing – by examining
the white matter tracts of participants' brains. These tracts are involved
in processing speed and the ability to understand and react to new
information. Brain scans and tests of thinking skills were administered
at the beginning and end of the six months to a group of 174 healthy
adults aged 60–79 with no cognitive deficits.

The only group in which white matter increased over the six months
was the dancing group, a result which, the researchers agreed, was
most probably due to the demands on the brain of learning complex
choreography. In all the other groups – which did walking, walking plus
nutrition, and active control (stretching and toning) – it declined, which
in itself is interesting as it shows that age-related decline is detectable over

as short a period as six months. The changes in white matter in the dancing group were also found to be related to increased processing speed. It should be noted here that similar positive changes in white matter have also been observed in young adults as a result of cognitive training.

Dancing is what psychologists call a 'complex intervention'. It's socially and emotionally rewarding, and it requires the integration of a range of brain activities – sensory and motor skills as well as thinking skills such as attention, decision-making and memory. Research is full of evidence that multi-component interventions are better than single ones, such as better nutrition alone or aerobic activity alone. We saw this in the animal experiments described earlier in this chapter, where exercise plus mental training proved a highly effective combination. Dancing has been described as the perfect human analogue of such 'combined' training.

Finally, like learning a language, taking up dancing is thought to buffer our brain against ageing. This discovery was first made in 2003 by the Albert Einstein College of Medicine in New York. In an elaborate study, they compared a range of activities such as reading, writing, doing crossword puzzles, playing cards, playing musical instruments, walking, tennis, swimming, golf – and dancing. All the cognitive activities (e.g. crosswords, card games) were associated with a lower risk of dementia – but of the physical activities, only dancing was. Moreover, dancing offered twice as big a reduction in risk of dementia – a reduction of 76 per cent – as any of the cognitive leisure activities, while swimming and cycling were not associated with a lower risk of dementia in this investigation! Why should dancing be so effective? It appears that free-form dancing requires constant, split-second decision-making, forcing the brain to rewire its circuits. This rewiring creates neuroplasticity and the resistance to brain damage (discussed in chapter 6) that scientists call 'cognitive reserve'.

Later research has produced even better evidence. In 2017, researchers at the Otto von Guericke University in Magdeburg came up with a cunning plan. They devised a devilishly difficult programme of dance training in which 'our elderly participants [aged 63–80] constantly had to learn novel and increasingly difficult choreographies'.[2] Over six long months, day in, day out. They compared the results for this group with those for another group who followed a programme of physical training with the same aerobic demand. What did they find? Unsurprisingly, both groups were fitter at the end of the experiment. However, dancing led to larger increases than other physical activity in hippocampus volume and four more brain areas, while only dancing produced increases in BDNF. There were no significant differences in mental performance between the groups; this the researchers, echoing other research groups, explained on the basis of time. Six months is long enough for the physical changes to the brain to occur, but not long enough to establish improvements in performance. Without hesitation, the researchers recommended their 'challenging dance program as an effective measure to counteract detrimental effects of aging on the brain'.[3]

Cognitive reserve

We have already come across the concept of 'cognitive reserve' in chapter 6. To reiterate, we can think of cognitive reserve as our brain's ability to improvise and find alternative ways of coping with life's big challenges. It's a concept that helps us to answer the question: why do some people manage to preserve their mental faculties well and others do not? This question is perhaps posed most acutely by the cases of some older adults who have the full array of pathological changes seen in the brains of Alzheimer's patients but have *no symptoms of the disease*. It is thought that we can protect our brain power better throughout our life if we build up our cognitive reserve. How do we do this?

Observational studies looking at people across their lives have shown that good cognitive reserve is related to educational and occupational achievement – and to the lifestyle we choose, including our leisure activities and our social life, as discussed in chapter 6. Good cognitive reserve is all about building up the neural networks in our brain. These networks change and adapt to the demands of everyday life through what we call 'brain plasticity'. The more complex and sophisticated the networks are, the better our cognitive reserve. Now, the good news is that cognitively stimulating activities will enhance our cognitive reserve and help us to cope with both physical trauma (such as concussion or stroke) and age-related biological changes in the brain, helping us to resist cognitive decline.

To summarize the findings so far, although we do not – yet – fully

Violin Concerto

understand the underlying mechanisms involved in cognitive stimulation, the secret appears to be to *challenge your brain*. With practice, in any form of mental and physical training, we might improve our performance. However, unless we really challenge the brain, the chances are we merely reinforce existing brain circuits. The brain will only rewire when it has to. If it doesn't need to, then it won't! The Global Council on Brain Health recommends that we incorporate enjoyable cognitively stimulating activities as part of a healthy lifestyle to help maintain our brain health and reduce the risk of cognitive decline as we age. They comment:

> Engaging in more cognitively stimulating activities is not necessarily
> better. The quality of the activities (including novelty, variety, level of
> engagement, cognitive challenge imposed and degree of enjoyment)
> is important. Moreover, the duration of time in which you spend

doing the activity plays an important role in the extent to which those activities will maintain or improve your brain function.[4]

A summary of what we know about the benefits of cognitively stimulating activities is shown in box 8.2.

BOX 8.2: COGNITIVELY STIMULATING ACTIVITIES – WHAT THEY ARE AND WHAT THEY CAN DO

- Cognitively stimulating activities are novel, mentally engaging activities, training or exercises that challenge our ability to think.

- Examples are learning to play a musical instrument, taking up a second language, having dancing lessons, learning to play cards, and learning complex, new mental and physical skills, like tai chi or juggling.

- To be cognitively stimulating, activities must be demanding, require concentration of effort, be intensely engaging, and require sustained and repeated practice over a prolonged period of time.

- They have been shown to improve our thinking skills, including attention, memory, decision-making and reasoning.

- They are thought to improve neuroplasticity and prompt 'rewiring' of the brain's neural networks.

- Learning a language and being bilingual have been shown to increase the size of areas of the brain such as the hippocampus, a vital centre for memory and learning; increase speed of processing; increase communication between the two halves of the brain; and buffer the brain against ageing.

- Dancing offers benefits to the brain along with the numerous personal, social and emotional benefits. It increases the volume and number of fibres in the white matter tracts, so improving processing speed, and reduces the risk of cognitive decline, including dementia in later life.

- Cognitively stimulating activities may enhance 'cognitive reserve', helping us to cope with challenging events in life, such as physical trauma or age-related cognitive decline.

- The brain can benefit from cognitively stimulating activities *at any age*; it is never too late to take up new activities that challenge the brain.

- There is increasing evidence that cognitively stimulating activities will reduce the risk of cognitive impairment.

- Cognitive decline may accelerate if we stop engaging in brain-stimulating activities.

Cognitive training

Can we train our minds to be sharper, faster and better, and conserve our brain power as we get older? Scientists have been researching cognitive training (CT) for about a hundred years, with mixed results. Cognitive training (or 'brain training') is any specific learning programme based on psychological principles with the defined aim of improving thinking skills – as opposed to the 'cognitively stimulating activities' we have just been considering, which are leisure or other activities with cognitive benefits. The first ever CT study, in 1910, attempted to train college students in memorizing lists of unrelated letters of the alphabet, and achieved some improvement but no general benefits outside the task – in some ways establishing the trend of later work. CT aims to improve abilities such as problem-solving, reasoning, attention, decision-making and working memory. It's based on the idea of 'neuroplasticity', which we've already met in several contexts: the brain's ability to respond and adapt to changing situations and environments. Neuroplasticity is known to decline with age.

Is CT effective? In general, we can say the following:

- There's a lot of evidence that CT improves performance at the task in hand.

- There's less evidence for a transfer of improvement to closely related tasks ('near transfer').

- And there's even less evidence that there's any transfer to distantly related tasks ('far transfer').

- Finally, though some studies report some benefits of CT in slowing down cognitive decline, there isn't any conclusive evidence that it delays or prevents the onset of cognitive impairment such as dementia or Alzheimer's disease.

But what about crosswords, brain-teasers and number problems? Surely they keep us sharp and improve our brain power? A number of interesting research projects have thrown some light on these questions. In 2019, the University of Exeter Medical School published the results of a study undertaken as part of the PROTECT research programme, run by the university and King's College London in collaboration with the NHS with the aim of understanding how healthy brains age and why some people develop dementia. This study involved over 19,000 adults aged over 50 at the outset, who were tracked for twenty-five years. It was found that the more regularly they engaged with puzzles such as crosswords and sudoku, the better they performed on tasks assessing attention, memory and reasoning, especially in terms of speed. One fascinating finding was that regular players appeared *eight years younger* in their performance than those who did not do such puzzles. Fine, you may say: to slow down my brain ageing and maintain my mental performance, crosswords are the answer. Not so fast. The study was 'correlational': that is, it looked only at the *association* between performance and 'brain games' and did not show *cause and effect*. So an alternative explanation for the results might be that the participants who played the brain games were the ones who were sharp-witted anyway – a phenomenon called 'reverse causation'.

There's also a lack of any very strong evidence that playing games like these transfers a benefit to everyday cognitive skills, which are inherently very complex. Another interesting study, this time in Germany, looked at jigsaw puzzles. The researchers found that when we do a jigsaw, we use certain vital skills such as perception, speed of reaction, flexibility, working memory and reasoning. A group of adults who did intensive jigsaw puzzling for thirty consecutive days improved in the specific skill of doing a jigsaw but not in 'global visuospatial cognition', compared to a control group (with no jigsaw training). In other words, the individual cognitive skills improved, but only within the framework of that particular game.

Will cognitive training make us more intelligent? Some fifty years ago, in a controversial paper published in the *Harvard Educational Review*, Arthur Jensen, a professor at the University of California, Berkeley, argued that IQ – the typical measure of fluid intelligence at the time – was largely immutable and resistant to environmental change. It began with what became a notorious statement: 'Compensatory education has been tried and apparently has failed.' The paper sparked a huge furore, political and academic, in the USA and beyond. After decades of ambiguous results, his view was put to the sword in 2008 in a landmark study by Suzanne Jaeggi and colleagues at the University of Michigan. They claimed their findings showed that fluid intelligence could in fact be improved, and that this was true for all levels of starting intelligence. What's more, they argued, there was a 'dose effect': that is, the more training, the more the gain. The Jaeggi study was predicated on an insight, new at the time, that *working memory* would be a key determining factor. The team used two validated tests of working memory. Since 2008, attempts to replicate these findings have generated mixed results, so the initial claims have been tempered. We do know now that fluid intelligence declines with age, albeit with big differences among individuals; but even if fluid intelligence can be altered by training, we still don't know how, if at all, this improvement will translate into our everyday lives.

The balance of evidence has in fact shown that, although performance on specific training tasks can be dramatically improved, *the transfer of this learning to other tasks, such as to have a generalized effect on our everyday lives, is poor.* The claim that training our working memory is the key to producing diverse beneficial effects was finally scotched by a 'meta-review' of eighty-seven scientific studies in 2016. The three authors concluded that training programmes involving working memory appeared only to produce short-term, specific training effects *that do not transfer to 'real-world', everyday cognitive abilities.* In other words, if we train our memory, we will end up remembering things better; but that won't make us more intelligent, more attentive or faster in our thinking.

Mind games

Strategy games are not new. They are as old as humanity. The fourth-century Chinese game of GO, Chaturanga in India, rudimentary chess in ancient Persia – all have the common elements of intellectual and strategic challenge. Later came the added visual, then tactile, challenges posed by jigsaw puzzles and the Rubik's Cube. More recently still, video and computer games have introduced a new level of complexity, spicing up the endless struggle to win.

The Last of Us. Grand Theft Auto. Sniper Elite. Thrilling, gripping, addictive. Testing our mental speed, manipulation and cunning. Firing up multiple networks in the brain – motor, sensory, visual – and lighting up its reward centres. Adrenaline, cortisol, serotonin all flooding the body, filling us with unbearable tension and shaping our behaviour, just as if we were small animals in a maze. And we solve the game! But what is it doing to our brain health? And to our capacity to think, reason and remember?

The latest science is clear: there is both an upside and a downside to gaming (gamers will be relieved to know that the balance is in favour of the upside). It's only about twenty years since researchers began working in this area, so the science is still in its infancy; but it is now of increasing importance as gaming technology becomes ever more widespread and the numbers of both 'casual' and 'dedicated' gamers increase (in all, some 32 million in Britain in 2017, making the UK the fifth largest games market in the world). In 2017, the first ever review of over a hundred studies on gaming and the brain was published. All of them showed that playing video games alters brain structure and function – and, moreover, our behaviour – across all ages from childhood to old age.

So what is the upside? First, video gaming improves our attention. It seems that gamers can pay attention for longer and have better 'select-ive' attention (singling out one item from another). Unsurprisingly, the brain regions involved in attention are also more efficient in gamers, 'switching on' more easily for demanding tasks. There are also what are known as 'visuospatial' benefits (abilities to perceive, recognize and manipulate what we see). The areas of the brain involved in visuospatial activity typically have more grey matter in gamers and also are more efficient, responding more quickly to stimuli. New research is showing that playing video games alters the electrical activity of the brain. In some areas, 'theta' waves increase, indicating a state of relaxation, but in other areas 'alpha' waves increase, indicating that one part of the brain is being inhibited in favour of another. Interestingly, those players who had more alpha waves were the ones who improved most in reac-tion time and working memory. Finally, the parts of the brain involved in control, for example rapid decision-making, become more active in gamers. However, studies have found that executive functions trained in a video game do not transfer well to other activities compared with those developed in other forms of cognitive training. It's as if playing the game endlessly rehearses and improves just those pathways in the brain used to become good at it.

And the downside? There is some evidence of a risk that video games may become addictive, leading to an affliction called IGD (internet gaming disorder). Researchers have found functional and structural changes in the brain's reward system by exposing addicts to 'gaming cues' then monitoring their brain responses. These changes are basically the same as those seen in other addictive disorders.

In summary, it seems that while video gaming produces specific benefits related to the particular requirements of the game, those benefits don't transfer well into everyday life. But what about other computerized games or puzzles which claim specifically to improve our brain power?

In 2014, two groups of eminent neuroscientists made public their sharply differing views of electronic or computerized brain games, a growth industry whose annual revenue had just for the first time surpassed $1 billion. One group, led by the Stanford Center on Longevity and the Max Planck Institute for Human Development in Berlin, published their conclusions that October. They didn't mince their words, stating unequivocally that the claims made by those promoting brain games were frequently exaggerated and at times misleading. In their summary, they stated: 'We object to the claim that brain games offer consumers a scientifically grounded avenue to reduce or reverse cognitive decline.'[5] The response was not long in coming. An open letter signed by over one hundred leading neuroscientists and researchers opined: 'We cannot agree with the part of your statement that says "there is no compelling scientific evidence" that brain exercises "offer consumers a scientifically grounded avenue to reduce or reverse cognitive decline".'[6] The letter was published on the Cognitive Training Data website under the auspices of Dr Michael Mezernich, Chief Scientific Officer of the Posit Science Corporation, a leader in the brain game industry alongside Lumos Labs, Inc.

So here we have a huge conflict: two consensus statements, signed by large numbers of equally eminent scientists, reflecting two opposing conclusions. What is the ordinary member of the public to believe? The first step in resolving the dilemma was the publication in 2016 of a review of 130 studies by Dr Daniel Simons and six of his colleagues at the University of Illinois. Their conclusion? They found little evidence that 'brain training' enhances performance on tasks unrelated to the specific activity of the game, or that training on brain games improves everyday cognitive performance. A major study in the UK led by Cambridge University and published in *Nature* had earlier reached similar conclusions – four years before the clash between the Stanford Center and Cognitive Training Data. In a six-week study, over eleven thousand adults were recruited, all of whom took a starting test of reasoning, memory and learning, to generate a benchmark of performance. They then trained regularly each week online – one group on tasks emphasizing reasoning, planning and problem-solving; a second group on 'game-like' tests of memory, attention and maths; and a third group simply used the internet to answer 'obscure' questions. Each group improved in what they were doing, but that was not the central question. When each group was then tested on the original psychological test, there was no difference between the groups, and the participants did not improve significantly on their original scores. The authors stated plainly: 'These results provide no evidence for any generalized improvements in cognitive function following "brain training" in a large sample of healthy adults . . . [the data] provided evidence that training-related improvements may not even generalise to other tasks that tap similar cognitive functions.'[7] A meta-analysis of twenty-three studies carried out at the University of Oslo in 2013 also confirmed these findings. It concluded that brain training did generate short-term, highly specific improvements in performing the task at hand, but produced no generalizable improvements in overall intelligence, memory, attention or any other cognitive ability.

That, however, was not the end of the controversy. The claims of the brain game industry inevitably drew the attention of the US Federal Trade Commission. In 2017, Lumos Labs agreed to pay a $2 million fine to settle FTC charges alleging that they had deceived consumers with unfounded claims that their Lumosity games raised everyday mental performance and reduced or delayed age-related cognitive impairment. Even so, a lot of people are clearly still enthusiastic about these games. In 2019, in an article entitled 'An evidence deficit haunts the billion-dollar brain training industry', the *Financial Times* forecast that by 2021 the brain games industry would be worth some $8 billion.

In conclusion, it's worth citing the consensus statement issued by the Global Council on Brain Health:

Most commercial products marketed as 'brain games' are not what the GCBH means when discussing the benefits of cognitive training.

If people play a 'brain game', they may get better at that game, but improvements in game performance have not yet been shown to convincingly result in improvements in people's daily cognitive abilities. There is insufficient evidence that improvements in game performance will improve people's overall functioning in everyday life. For example, we do not have evidence establishing that getting better at playing Sudoku will help you manage your finances any better.[8]

A summary of what we do know from the evidence about cognitive training is given in box 8.3.

BOX 8.3: WHAT WE KNOW ABOUT COGNITIVE TRAINING

- Research shows that specific skills such as processing speed, working memory or attention can be improved by specific training, for example in video games or leisure activities like puzzles and 'brain-teasers'.

- In these activities, numerous cognitive skills are engaged, including perception, processing speed, flexibility, working memory, reasoning and episodic memory, and these skills improve for the purpose of improving how well we do in the game.

- However, there is little or no evidence that any of these benefits transfer into general mental ability – in other words, there's no 'far transfer'.

- Regular engagement with cognitive-based activities, like crosswords and sudoku, is associated with better memory, attention, speed and reasoning, but may not be the cause of these improvements.

- Computerized brain games improve the skills used in mastering the game, but there is no conclusive evidence that they improve everyday cognitive abilities, or that they slow down cognitive decline or reduce the risk of dementia.

Staying sharp – what to do and how to do it

On the basis of the latest evidence we can offer the following practical advice:

- Challenge your brain. Take up a stimulating activity which is difficult, requires concentration and takes time to master. An activity which is easy and habitual, and in which you 'cruise', won't work!

- Examples of good, brain-stimulating activities include learning a new language, dancing, card games, chess, and challenging forms of physical exercise such as tai chi, yoga and juggling – but there are lots of others, such as learning new technologies, creative writing, making art and community volunteering. We don't have good evidence that any one particular cognitively stimulating activity is more effective than another for maintaining your brain health.

- Incorporate these activities into a generally healthy lifestyle to protect your brain against age-related change. Remember that taking up and regularly practising a demanding form of physical exercise, such as tennis or even bowls, is especially good for the brain.

- While you should start as early in life as you can, it's never too late to take up new activities, and it's important to maintain learning activities throughout life. Don't let age limit the scope of your activities or your intellectual life.

- Choose something that you enjoy – willpower alone will not get you through. Have a partner or be a member of a group to help you to stay with it.

- Do jigsaws and other challenging puzzles, including crosswords, sudoku, maths games and other 'brain-teasers' if you enjoy them. They won't necessarily improve your general thinking skills, but the specific skills you need to get good at them will improve – and they won't do any harm!

- The same is true of computerized 'brain games' – play them if they are enjoyable, but don't rely on them to improve your general thinking skills or to slow down age-related changes in the brain. Nor is it at all likely that they will protect you against Alzheimer's or other neurodegenerative changes.

If we were the mice in the maze, and knew what the scientist in the lab coat knew, we would tackle the maze with gusto, challenging ourselves and working with all our mind, tiny though it would be, to stimulate it as much as we could. There is no 'magic cure' for maintaining and saving our brain health. The idea that there is some 'silver bullet' activity which will boost your brain power is a seductive one. But it's a fallacy. Even so, thanks to science, we know that whatever activity we choose, *if we challenge our minds and stimulate the brain to learn new things*, we can get pretty close.

To sleep or not to sleep?

IN 1963, AS THE Beach Boys were playing on the radio and Christmas was approaching, two California schoolboys threw a coin. They were deciding who would be the guinea pig in a school science project they had designed – to beat the world record for staying awake. The lucky 'winner' was Randy Gardner, a 16-year-old from San Diego. When the experiment was over, he had stayed awake for eleven days and twenty-five minutes. It yielded some fundamentally important observations, fortunately recorded by William Dement, one of America's few sleep researchers at the time. Nearly forty years later, Gardner still holds the world record – which is unlikely to be broken, as the *Guinness World Records* will no longer accept entries. Why? *It is much too dangerous for the brain.*

There is no more intractable health problem in modern life than sleeplessness. Insomnia, difficulty sleeping and sleep disorders are all prevalent in today's world. It is as if we are all in some ghastly sleep deprivation experiment. Shift work, long commuting hours, caffeine, stress, social life, travel, technology and, as we get older, age-related changes all influence our sleeping habits.

Lack of sleep won't kill you, but it could come close. It will seep into every nook and cranny of your being and corrode your health, your well-being, your work and your leisure. Sleep is not an add-on luxury in our lives. It is a 'non-negotiable biological necessity', according to neuro-scientist Matthew Walker, author of *Why We Sleep*. And if we can't sleep, the reasons why are in our heads. In our brains. But before we go there, history has important lessons to teach us.

The terrible beginnings of sleep science

In 1907, two French scientists conducted a distressing experiment. Taking two healthy dogs, René Legendre and Henri Piéron tied their collars to a wall so that the animals were unable to sit or lie down. For ten days. A truly terrible experience of sleepless imprisonment for the hapless canines. After ten days of this agony, they were euthanized and their cerebro-spinal fluid (bathing the brain and spinal cord) was extracted, then injected into healthy, active dogs. Within one hour they were asleep. The conclusion? There was a mysterious, sleep-inducing molecule which Piéron and Legendre called 'hypnotoxin' – but of which they knew nothing.

These two were not the first to recognize the deadly power of sleep deprivation. Ancient documents reveal that, among others, the early Chinese understood only too well its lethal effects. Forced insomnia was used not only as a form of excruciating torture but also as a brutal means of execution. As Marie de Manaceine, Europe's first female doctor and a physician at the Military Academy of Moscow, concluded, 'the total absence of sleep is more fatal . . . than the total absence of food'.[1]

Legendre and Piéron were pioneers in what became a rich tradition of research into sleep – most of it, fortunately, not so gruesome, but all of it contributing to the gradual revelation of sleep's secrets. Through it, we have learned much about our human need for sleep, its importance, how it is controlled, and why and how it goes wrong (sleep disorders). In essence, sleep is vital to human health; but we are only just beginning to realize the extent of its impact on our lives – and vice versa. The assertion of this reciprocal association is based on the emergence relatively recently – only fifty years ago – of widespread sleep problems, sparked by a technological revolution which has led to today's 24-hour society and huge changes in our lifestyles. It's noteworthy that we Brits are the most sleep-deprived society in the world, with 37 per cent of us feeling we do not get enough.

The mystery of sleep

Despite all the research that has been done over the century and more since 1907, there are still big gaps in our knowledge. I shall highlight one of these, the evolution of sleep, by introducing you to the Mysterious Dolphin. When I was a graduate student, my sleep professor reduced my class to silence by asking a simple question: 'How do aquatic mammals sleep, if they need to remain awake to breathe?' No one could answer it. It seemed to be an unresolvable paradox. The answer is as ingenious as it is intriguing. Only one half of the dolphin's brain sleeps at a time, with the opposite eye closed. The message I drew from this was that sleep was of critical importance. But the reverse questions from the students were equally difficult to answer: Why did our dolphin need to sleep anyway? What is sleep's biological purpose? Why do all mammals do it? How did it evolve? My professor had some answers, but could offer no certainty. The biological purpose of sleep (but not its physiology) remains largely a mystery. But the question is of such importance in understanding sleep in humans that we should pursue it in a bit more detail.

Biologically, evolution is about survival – reducing the risks of disease or death at the hands of predators, climate and other adversities, so that the 'fittest' survive long enough to reproduce, thereby passing on the most successful adaptations. Sleep appears to be incompatible with survival – it prevents feeding and procreation, and could expose the sleeper to attack by predators. Conversely, it has been argued that sleep may reduce the risk of predation by enforcing behaviour modifications such as living in social groups. Some interesting research into the Hadza people in northern Tanzania found frequent night-time waking and widely differing sleep schedules. Over a three-week period, there were only eighteen minutes when all thirty-three tribe members were asleep together. The scientists behind the work concluded that fitful sleep could be an ancient survival mechanism designed to guard against nocturnal threats. It seems most likely that sleep evolved to ensure that species are not active when they are most vulnerable to predation and when their food supply is scarce – that is, at night. This principle may well be the driver behind our circadian or 24-hour rhythm, seen increasingly as one of the key factors in human health. We will look at this rhythm later on in this chapter.

One of the major trends of evolution is the development in certain groups of mammals – including us – of bigger and bigger brains with more and more complex structure and function. In the case of humans, this progressive evolution is associated with standing upright, the prehensile grip (opposable thumb), working in social groups, suppression of needless emotion – that is, emotion that does not contribute to primary survival processes such as planning and executing the killing of prey or rivals – and advanced planning and decision-making. The bigger brain means that the infant child has to be nurtured for a long period to allow the brain the time it needs to develop, and this requires intimate bonding and pairing between the parents. Surprising as it may sound, the evolution of our complex advanced brain is the primary reason why humans have sex face to face – pretty much the only primates to do so. Why should this be? Because the nine months the human infant spends in the womb isn't a long enough time to allow the brain to mature fully. So this process has to carry on after birth, when the vulnerable child and nursing mother need protection. Eye-to-eye sexual contact promotes bonding, encouraging the male to stay around for far longer than in any other primate.

Given what we now know about the brain and sleep, it is not unreasonable to argue that the bigger brain is responsible for improvements in our survival capacity, and, conversely, that longer, more complex and more dynamic sleep is required for that bigger and more advanced brain to function effectively. Sleep must, after all, confer some pretty substantial benefits to outweigh the risks entailed. As sleep science pioneer Allan

Rechtschaffen put it: 'If sleep does not serve an absolutely vital function, then it is the biggest mistake the evolutionary process ever made.'[2]

A happy accident

To understand those benefits, and how we came to know about sleep, we must turn the clock back to 1893, and our eyes to Lower Saxony, where Hans Berger was a young cavalry officer on a training exercise with the Prussian army. Growing up in an aristocratic family, he had dropped out of his maths degree after a single semester at the University of Jena, one of Germany's oldest and most prestigious academic institutions, scholarly home to many notable figures, including Schiller, Hegel, Schlegel and Haeckel. Now, far from the rarefied air of the university cloisters, bedecked in the rich purple and red of his regiment, he cut a fine figure on horseback. Then, in a split second, disaster struck. Startled, his horse unexpectedly reared, throwing Berger into the path of an oncoming gun carriage. Fortunately for the young officer, the artillery driver reacted with impressive speed, expertly swerving around him; shaken but unhurt, Berger got to his feet, mesmerized by his miraculous escape.

So far, so unremarkable: but at that precise moment, many miles away, Berger's sister sat up in alarm, overwhelmed by a feeling that her brother was in mortal danger. She implored her father, at much inconvenience, to telegram his regiment. As the sequence of events became clear, everyone involved was astonished at the extraordinary coincidence. Many years later, Berger claimed that 'it was a case of spontaneous telepathy in which at a time of mortal danger, and as I contemplated certain death, I transmitted my thoughts, while my sister, who was particularly close to me, acted as the receiver'. His military service completed, Berger resumed his studies at Jena – but no longer in mathematics. Obsessed by the idea of how his mind could have transmitted a signal to his sister, Berger took up the study of medicine, with the goal of discovering the physiological basis of 'psychic energy'. His central theme became 'the search for the correlation between objective activity in the brain and subjective psychic phenomena'.[3] This goal remained elusive; but some three decades later, Berger was responsible for a huge advance in brain science when he recorded the first human electroencephalogram (EEG), monitoring electrical activity in the brain. This milestone event, met with derision and incredulity at the time, became the foundation of modern sleep studies. In 1953, at the University of Chicago, Eugene Aserinsky and Nathaniel Kleitman discovered rapid eye movement (REM) sleep using EEG techniques. Sleep research was born – and with it, our understanding of sleep's true nature and benefits.

The nature of sleep

Sleeping is an integral – and, as later twentieth-century research has revealed, incredibly complex – part of our lives. It would be easy to think that when we sleep, the 'lights go out' and the brain is resting. Nothing could be further from the truth. The brain (as we will see later) is incredibly active during sleep. Sleep is not a period of biological inactivity between periods of being awake. Further, EEG studies have revealed

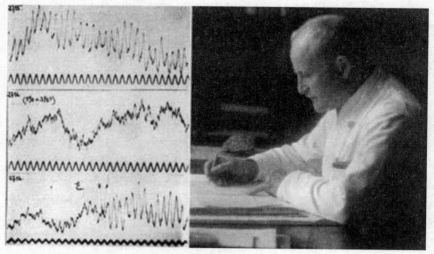

Hans Berger and his early EEG recordings

that all sleep is *not* the same. The brain generates two distinct types of sleep – non-REM (NREM) sleep, which includes slow-wave sleep (otherwise known as 'deep sleep'), and REM sleep, also called 'dreaming sleep'. When we are sleeping, our brain will cycle repeatedly through the two different types.

The first part of the cycle is NREM sleep, which in turn breaks down into three stages (labelled N1, N2 and N3). The first stage is that drowsiness between being awake and falling asleep with which we are all familiar. The second is light sleep, when our heart rate and breathing level out and body temperature drops. The third stage is deep sleep (slow-wave sleep): this accounts for most of the sleeping we do and is characterized by large, slow brain waves, relaxed muscles and slow, deep breathing. Scientists now have concrete evidence that two groups of cells – one in the hypothalamus and one in the brain stem – are involved in prompting deep sleep. When these cells switch on, they trigger a loss of consciousness. Though REM sleep was previously believed to be the most important sleep phase for supporting learning and memory, newer data suggest that in fact NREM sleep is more important for this purpose, as well as being the more restful and restorative phase of sleep.

As we cycle into REM sleep, our eyes move rapidly behind closed lids, and brain waves (detectable by EEG) are similar to those during wakefulness. Our breathing rate increases and the body becomes temporarily paralysed as we dream. This mode of sleep is bizarre: the dreamer's brain becomes highly active while the body's muscles are immobilized and breathing and heart rate become erratic. The purpose of REM sleep remains a biological mystery, despite our growing understanding of its biochemistry and neurobiology. But we do know that a small group of cells in the brain stem, called the subcoeruleus nucleus, controls REM sleep. When these cells become injured or diseased, people do not experience the muscle paralysis associated with REM sleep, and this can lead to REM sleep behaviour disorder – a serious condition in which people act out their dreams violently.

The NREM/REM cycle then repeats itself – but with each cycle we spend less time in stage N3 and more time in REM sleep. On a typical night, we'll cycle four or five times. The reasons for these alternations are as yet unknown to science.

Why we need sleep

To get to the bottom of why we need sleep, we need more than anecdotal evidence. It is one thing for science to describe what is happening in the brain during sleep. It is quite another to unravel why this mysterious sequence of processes is so important. Thanks to the considerable weight of research currently dedicated to the subject (two hundred sleep labs in the USA, ten world-leading centres in UK universities and over twenty scientific journals), we are beginning to get a much clearer idea of why the brain needs sleep, though we still lack substantial empirical evidence to back up some of the explanations being proposed.

When we quit for the day, as the saying goes, our brain does some serious work. The evidence to date supports the following reasons why we sleep:

- sleep helps to forge new synapses (connections between nerve cells) and solidify memories, experiences and emotions;

- sleep allows the brain to filter out unimportant synapses and thereby to prevent overload in the higher parts of the brain (the cortex);

- sleep de-toxifies the brain, removing unwanted cellular detritus and potentially damaging protein molecules (such as beta amyloid which, as we saw in an earlier chapter, is associated with dementia) through the recently identified 'glymphatic system' – a special drainage mechanism which operates twice or three times as fast during sleep as when we are awake;

- sleep may also help the brain repair itself by removing waste products and so-called 'free radicals', which cause 'oxidative stress' in the brain cells, thereby acting as the brain's garbage disposal system.

Sleep affects almost every organ and tissue in the body, including the cardiovascular system, the metabolism and the immune system as well as the brain and nervous system. We know beyond doubt that poor-quality sleep increases the risk of a range of disorders including high blood pressure, cardiovascular disease, diabetes, depression and obesity – all associated with shortened lives. We hear a lot about the importance of anti-oxidants, which have cropped up in several chapters of this book already – and sleep is relevant to them too. Lack of sleep reduces the anti-oxidant defences of the body, inhibiting the ability to remove molecules that promote cell damage and inflammation, such as free radicals and oxygen-reactive substances.

In summary, sleep is of critical important in re-energizing our body's cells, clearing waste from the brain, and supporting learning and memory. It also plays vital roles in regulating mood, appetite and libido.

However, the ways in which sleep delivers these effects are largely unknown. What is particularly fascinating is that molecules and meta-bolic pathways only found *outside* the brain seem to be influenced by sleep – and then the effects are seen *in* the brain. We are still at the beginning of a long, long research journey to unravel sleep's secrets. But we already know enough to make our lives better.

Dozing off at an inopportune moment can be embarrassing or even dangerous. And never more dangerous than on 28 January 1986, when the space shuttle *Challenger* broke apart seventy-three seconds into its flight, killing all seven crew members. The principal cause of the accident is widely known: the failure of the so-called 'O' rings, which allowed pressurized burning gas to escape. Less well known is the role played by operational decisions made during a conference call the day before, when critically important managers had had less than two hours' sleep and had been on duty since one o'clock that morning. The Presidential Commission on the Space Shuttle *Challenger* Accident cited human error and poor judgement related to sleep loss as a pivotal contributing factor – so significant, in fact, that a new policy on management decision-making for future launches was implemented. Sleep loss as a pivotal contributing factor can be laid at the door of other devastating industrial accidents, including those at the nuclear plants at Three Mile Island in 1979 and Chernobyl in 1986. Difficult though it is to gather all the evidence, it is known that in both those cases, human error in the sleepless early morning hours was implicated.

We should be grateful that such incidents do not occur more frequently; and for that we have to thank the merciful intervention of two bodily systems whose interaction reliably governs our pattern of waking and sleeping. They are the *circadian rhythm* and the *homeostatic sleep drive*.

Nearly fifty years ago, in a darkened cave in Texas, a perilous experiment took place which would dramatically change what we know about our daily rhythms and sleep. In 1972, French geologist Michel Siffre entered the forbidding Midnight Cave, 100 feet below ground, where he planned to remain for six months. As a researcher, I can only imagine he did it alone because he would never have received ethical permission to involve others (the history of science is littered with examples of such awesomely impressive go-it-alone individuals who have risked their necks for science).

At the time he began his subterranean existence, alone and with no natural light or sound, we already knew that the normal physiology of the body runs on the basis of an internal clock plugged into the external environment. Such rhythms were known to the ancients well over two thousand years ago, though they had little idea, if any, of their significance. The Greek writer Androsthenes described how the tamarind tree opened its leaves during the day and closed them at night, daily movements conspicuous to the soldiers of Alexander the Great who first reported them during their conquest of Tylos (now

Bahrain). Two millennia later, in 1729, the French astronomer Jean-Jacques d'Ortous de Mairan showed that these rhythms actually had *an internal origin*: he recorded that the leaf movements of the sensitive heliotrope plant (probably a mimosa) persisted throughout the hours of darkness.

In the constant darkness of his self-imposed exile, our modern-day Frenchman recorded his own somewhat different responses, cut off from the natural cycle of day and night. He slept, rose and ate as he wished, and kept a written record of his activities. His records yielded some important findings. Up to this point, scientists had believed that the human natural daily or 'circadian' rhythm (from the Latin *circa diem* – about a day) was a 24-hour cycle, tied into day length. But Siffre disproved this. Separated from the natural cues of night and day, his circadian biorhythm free-ran, averaging between twenty-four and twenty-five hours, so that his periods of waking and sleeping became distorted in his light- and sound-deprived environment. Because the rhythm does not work on a cycle of precisely twenty-four hours, it has to be reset each day; this is done by means of daylight entering the brain via the eye (a point we shall revisit later in this chapter). Second, Siffre found that his sense of time evaporated. Without natural cycles of daylight and darkness, it became impossible for him to discern the passage of time accurately.

Such environments as that to which Siffre subjected himself are completely unnatural and, as it turned out, after as long as six months, immensely damaging to health. As he emerged from his underground ordeal, Siffre was found to be suffering from impaired movement and coordination and from memory problems – all evidence of the ill-effects of detachment from our brain's natural rhythms. From jet-lag to shift work to daylight saving, many of us will be familiar with disturbances in how we feel, how we perform and above all how we sleep when our internal clock becomes disturbed. Similar problems are encountered by astronauts living for long periods in space.

Staying in sync

So what is this all-important circadian rhythm? In essence, it is a cycle of arousal, during which our level of consciousness or wakefulness rises and falls over a period of about twenty-four hours. It is connected to other cycles regulating body temperature, hormone levels and cellular metabolism. Body temperature, for example, starts to rise at about 6 a.m., climbs quickly and peaks at about 12 noon, after which it plateaus and oscillates until about 10 p.m. (with a mini-peak at about 8 p.m. after eating), at which point it starts falling, continuing to decline gradually throughout the night. It is almost completely independent of external conditions, but it is synchronized with the level of arousal or wakefulness – which is one reason why it is difficult to sleep if the surroundings are too warm. As the body quickly cools down at night, melatonin, the 'sleep hormone', is released, provoking sleepiness. Paradoxically, there is some evidence that taking a warm shower or bath helps us to fall asleep. The explanation is twofold. First, a warm shower or bath increases our state of relaxation, lowering our arousal level; second, warming the body starts up its cooling mechanism, so that eventually core body temperature starts to fall. The evidence suggests that a bedroom temperature of about 18 degrees Celsius is optimal for sleep, though there are huge individual and even cultural differences. Box 9.1 sets out some ways in which you can help yourself to stay in sync with your circadian rhythm.

BOX 9.1: STAYING IN SYNC – PRACTICAL TIPS TO HELP YOU SLEEP

- Keep the bedroom at a temperature comfortable to you – 18 degrees Celsius is recommended.
- Allow the temperature to cool after you get into bed.
- Take a warm shower before bed.

- In the summer, leave a window open or use fans or air-conditioning as needed.
- Do not let the bedroom get very cold (less than 12 degrees Celsius), so avoid open windows in winter.
- Keep the bathroom warm (and close the window in the winter).
- Avoid eating or drinking less than three hours before going to bed.
- See your doctor if you have an 'over-active bladder' which keeps waking you at night.

Our circadian rhythm, illustrated in figure 9.1, results from the operation of an internal biological clock. This clock, which is responsible for regulating the vast number of daily rhythms underlying our bodily functions, is located in a relatively small collection of nerve cells deep within the brain. Our primary circadian clock is located in a small organ in the base of the brain which has already featured prominently in this book: the hypothalamus. This modestly sized structure is hugely important, being critical to the regulation of all physiological function

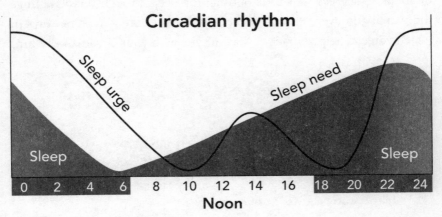

Figure 9.1: The circadian rhythm

– for example, it is the seat of fundamental appetitive motivations, such as hunger, thirst and sex drive. It is connected by nerve fibres to the controlling centre of our hormonal or endocrine system, an equally tiny gland called the pituitary which is often referred to as 'the leader of the endocrine orchestra'. Together they form the basis of a supremely important control system, the 'hypothalamic–pituitary axis' or HPA (already mentioned in chapter 4 in the context of the gut biome). In animal experiments, removal of the hypothalamic cells which make up the clock results in the complete loss of any regular sleep–wake rhythm. The clock receives information about light via the eyes, so that under normal conditions it is closely synchronized with our sleep–wake cycle. It is an 'alert' signal from this clock that keeps us awake, increasing in strength throughout the day.

It is now believed that harmonizing our daily activities with our natural rhythms is a vital factor in keeping our brains healthy across our lives. Eating at the 'wrong time' (e.g. two in the morning) or at different times on different days, or being active at night (e.g. staying up late to party or to write a work-related report), are thought to be very destabilizing to our rhythms, and to the control mechanisms in the brain. Fortunately, the effects will only persist as long as the disturbance. Restore eating to 'normal' hours and the 'de-sync' disappears. Our performance, both mental and physical, is greatly dependent upon the maintenance of our rhythm, and upon the point we are at within it. It has even been said that using circadian rhythms to estimate sporting performance will predict winners better than the bookies' odds. The evidence clearly shows that sleep of sufficient duration, continuity and intensity (depth), without disrupting the circadian rhythm, is necessary for high levels of attention and mental performance while we are awake – and to prevent physiological changes that promote ill-health, including in the brain.

The sleep advice arising from this evidence is correspondingly relatively straightforward. Regularity (what sleep experts call 'sleep hygiene') is the key. The general rule is to get up at the same time every day, seven days a week. If you sleep well, aim to go to bed at the same time every day too; if you have trouble sleeping, it is better to wait until you are drowsy or sleepy. This may mean getting into bed and waiting for that feeling of drowsiness (the N1 stage described earlier). How we feel at bedtime depends to a large extent on what we've been doing during the daytime – for example, whether we have napped or not, or whether we are physically tired. For this reason, the Global Council on Brain Health recommends taking regular physical activity as an aid to good sleep. If you do have trouble sleeping at night but often doze off in the evening (for example, when watching television), you should either 'listen to your body' and go to bed when you feel tired (and possibly adopt an earlier sleep schedule) – or, if you feel it is too early to go to bed, try keeping yourself more alert by standing up and being physically active.

Is it worth getting medication to help you sleep? The picture is mixed. Hypnotic drugs (such as Zolpidem) and sedatives do work, but GPs are generally reluctant to prescribe them and, if they do, will advise using them as only a temporary measure – for two reasons: one is to avoid drug dependency, and the other is that they become less effective with continual use. So if you are using prescription medications to help you sleep, consider limiting their use to three nights during the week, unless your doctor says otherwise. Some people use over-the-counter medications, but these can have side-effects and are best avoided, particularly as we get older. Dietary supplements for sleep, for example supplementary melatonin, may have benefits for some, but the scientific evidence on their effectiveness is inconclusive.

Box 9.2 offers some practical tips on how to improve your sleep hygiene.

BOX 9.2: PRACTICAL TIPS TO HELP YOUR SLEEP HYGIENE

- Be physically active during the day; being a couch potato will not help you to sleep!
- Keep the bedroom for sleeping (and sex) – avoid using it as an office, for hobbies or for watching television.

- Keep pets that disturb your sleep out of the bedroom.

- Keep a regular sleep schedule – go to bed and get up at the same time, every day, if at all possible.

- If you sleep well, stick to your regular times.

- If you don't, do something relaxing and clear your mind before you retire to bed; wait until you are drowsy or sleepy before going to bed; or get into bed and wait until you are drowsy before lying down to sleep.

- Go to bed earlier if you feel sleepy during the evening ('listen to your body').

- Get a comfortable bed, mattress and pillow.

- Avoid over-the-counter medicines if possible, and do not rely on them.

- See your doctor if you have persistent sleep problems, but try not to become reliant upon prescription medicines.

In simple terms, the argument runs like this: our brains require adequate sleep, and adequate sleep in turn depends on adequate sunlight to 'sync-in' our circadian rhythm. As darkness approaches, the 'sleep hormone' melatonin is produced and prepares us for sleep. That's the 'official' version. However, the messages about light and sleep are rather more subtle than this suggests. Light is medicine. But there is also 'junk light', in the same way that there is 'junk food'. It all comes down to the wavelength of light and when we are exposed to it. In our everyday lives, light is universal; unlike our ancient forebears, we have to actively seek out darkness. We live in a sea of artificial illumination – and it is rich in short-wavelength or 'blue' light. This short-wavelength light is especially strong in energy-saving light bulbs and in digital devices, such as laptops, tablets, mobile phones and television screens. In the mornings and early afternoons, natural blue-wavelength light is extremely beneficial – it raises alertness, reduces daytime sleepiness, improves reaction time and extends attention span. However, at night it is the assassin of good sleep. It is especially aggressive in keeping us awake. In a 2017 study in the Assuta Sleep Clinic in Israel, it was found that when healthy young adults were exposed to blue light from their computer screens between 9 p.m. and 11 p.m., it reduced their total sleep time, suppressed melatonin production and increased the number of times they woke in the night. Their body temperature didn't fall, they were more tired during the day and they had more negative moods. Sounds familiar? Conversely, exposure to red (long-wavelength) light during the same two-hour period of the evening did not interfere with sleep: body temperature fell and sleep progressed as usual. Why should this be? It works like this. Because of the scattering of sunlight by particles in the atmosphere (known as Rayleigh scattering), the colour of the sky depends on the position of the sun in relation to the horizon. So at midday the light is perceived as blue, and in the evening, when the descending sun is closer to the horizon, it is perceived as red. If you treat the brain to a dose of blue light, the brain reacts as if it's the middle of the day. Not a good idea if you're trying to go to sleep!

The message? Keep away from blue light at night – and get plenty of natural light during the day. Not only should we keep the bedroom quiet and dark at night, we should stop using devices such as tablets, mobile phones, laptops etc. at least two hours before we go to bed – and, ideally, keep smartphones, televisions, electronics and screens of any kind out of the bedroom. If you can't, use blue-light filters or screens to remove the harmful wavelengths. Similarly, if you have to get up in the night, use a soft night light rather than turning on the overhead light; and replace cool white or blue-coloured night lights with red or orange-coloured ones.

It is arguable that the price we are paying for our wanton relationship with technology is an epidemic of insomnia and sleeplessness. Across a lifetime of such poor habits, the consequences for good brain health do not bear thinking about. Box 9.3 summarizes the key points about light and sleep, with practical tips on how to manage your exposure to light at different times.

BOX 9.3: LIGHT AND SLEEP – PRACTICAL TIPS

- Expose yourself to natural daylight during the day.
- Keep away from blue light at night – it is the enemy of good sleep.
- Keep the bedroom quiet and dark at night.
- Use non-blue lights or bedside lamps, or blue-light filters.
- Keep screens of any kind – tablets, phones, laptops and televisions – out of the bedroom.
- Avoid using screens in the last hour before you go to bed – extend the time you spend getting ready for bed, to put more distance between screen use and sleep.

Staying in balance

Now we come to the second of our built-in sleep controls, the homeostatic sleep drive. We have all experienced that undeniable drive to sleep – nodding off in boring meetings, slumbering during a movie – or, frighteningly, dozing off at the wheel of a car. They all serve as unpleasant reminders of the power of the sleep drive. It is omnipresent, even though we may be unaware of it. It grows throughout the day and can only be assuaged by sleeping. It is what biologists call a 'homeostatic' mechanism, and is a prime example of the way the body sustains a constant 'internal environment'. Homeostasis is vital for life. If it fails, then so does the body. Other classic examples are regulation of glucose, body temperature and hydration.

In 1932 Joseph Barcroft, a British physiologist, presciently declared that higher brain functions required the most stable internal environment of any organ in the body. When body temperature falls, blood vessels constrict and we shiver; when blood sugar levels rise, the pancreas secretes insulin; *and when we remain awake for a long time, structures in the brain promote sleep.*

The longer we stay awake, the longer we restrict our sleep, the stronger the sleep drive becomes. We can't measure it directly, but neuroscientists think that the force of the sleep drive is determined by the level of brain activity while we are awake. Remember Legendre and Piéron's sleep molecule, which they called 'hypnotoxin'? One current candidate for the 'sleep molecule' is adenosine, a by-product of energy consumption by our cells that is said to play a significant role in promoting the sleep drive.

We can now attempt to explain how our alertness and sleepiness vary during a normal day. In the early morning, the alerting circadian signal starts to make itself felt, at a time when the sleep drive is very low. We begin to wake up. For most of us, our arousal levels rise throughout the morning from about 6 a.m., along with another regular rhythm – the rise in release of cortisol, a stress-related hormone which increases our level of alertness. Cortisol peaks at about 9 a.m. – which makes this, for most people, probably the best time to drink your early morning coffee or tea, thereby reinforcing the rise in alertness already stimulated by the hormone. If you drink your coffee any earlier, you may not be getting the full benefit of it. Cortisol falls rapidly after 9 a.m. and flattens off at about 1 p.m., so drinking coffee throughout the morning (between 9 a.m. and 12 noon) can, by contrast, be more beneficial. The caffeine in coffee blocks the action of adenosine and inhibits certain sleep-inducing hormones, promoting wakefulness and lengthening the circadian rhythm. This timetable depicts the average – there are, of course, early and late risers. The message is: examine your own rhythm and adapt the advice accordingly – the principles are the same, as is the shape of the cycle. It just starts rising at slightly different times. If you are an early riser, for example, you would want to give your cortisol level time to rise and then drink your morning coffee or tea, rather than drink it straight away on waking. And of course, individual habits and preferences differ.

It's important to understand how caffeine works. Within forty-five minutes of drinking a cup of coffee, all the caffeine in the liquid has entered the blood and is on its way to the brain. It easily passes the blood–brain barrier. In the brain, caffeine inhibits the action of adenosine, so that the sleep drive goes down, and also increases the numbers of receptors for several important neurotransmitters: serotonin (26–30 per cent increase), GABA (65 per cent increase) and acetylcholine (40–50 per cent increase). We become more alert. And this is why we feel upbeat and have more energy after we've had a coffee. Caffeine is 'cleaned up' by the liver and its 'half-life' is about five hours. so if you drink a cup of coffee containing 100 milligrams caffeine at 2 p.m., 50 milligrams of it will still be circulating at 7 p.m. So two cups of coffee after lunch will deliver 50 milligrams of caffeine at midnight – easily enough to keep you awake.

Now back to our day! By midday, the alerting signal peaks; and then, shortly after lunchtime at around 1 or 2 p.m., it 'hiccups'. The grogginess that many of us often experience in the early or middle afternoon, commonly attributed to a heavy lunch or a dull meeting, is usually the result of this brief lull in the strength of the alerting signal. As the sleep drive continues to climb, there is an hour or two each afternoon during which that signal fails to keep pace with our activity level and alertness often suffers. This is the ideal time for 'power napping', and though many cultures have incorporated this lull into their lives as a mid-afternoon siesta, the evidence tells us that we should not nap for more than about forty minutes. The main reason is that taking too long a nap reduces the sleep drive and so makes it harder to fall asleep at night.

In the early evening, we experience the combined effects of work-related fatigue, an increased sleep drive and a slow-down in the alert signal. We add to this potent cocktail of sleep induction if we drink alcohol, a depressant and mood-modifying drug which interferes with neurotransmitters – among other things, increasing the production of serotonin. Serotonin is often referred to as the 'happy drug' because it promotes feelings of well-being, and low levels have been linked to depression. It is also the precursor molecule for melatonin, a sleep-inducing hormone. Further, consuming alcohol reduces the behaviour-constraining effects of the cerebral frontal cortex – one too many drinks and our inhibitions 'go to the wall'. It is no wonder, then, that so many people go for an early evening drink. However, consumption of large quantities of alcohol floods the brain and its early effects begin to reverse. Alcohol may help to 'put out the light', but one of the many reasons why too much of it is a bad idea is that overconsumption often leads to fitful sleeping. These effects will vary with age, health condition and individual constitutional differences. But the general rule should be: avoid alcohol several hours before bedtime.

As with public guidance on sleep and caffeine, when the Global Council on Brain Health issued this advice it met with a storm of protest. Among the comments that flooded in were 'If it wasn't for alcohol, I would *never* get to sleep at night'; 'I swear by my nightly tipple'; and 'Britain is a nation of drinkers, do not try to stop us!' It is true that alcohol is a relaxant, generates a feeling of well-being via serotonin release in the brain and induces sleep – but all the evidence (including a new review of twenty-seven studies) shows that it also represses the normal sleep cycle and ruins sleep quality, as well as dehydrating the brain (if you need reminding how dangerous this is, flip back to chapter 5), with both short-term and long-term ill-effects on brain health. Key points on caffeine and alcohol consumption are set out in box 9.4.

BOX 9.4: ALCOHOL, COFFEE AND STIMULANTS – PRACTICAL TIPS

- Avoid heavy consumption of alcohol immediately before going to bed.
- Restrict evening drinking of alcohol to safe, daily limits for your age and sex.
- Avoid drinking coffee after 2 p.m.
- Tea contains much less caffeine than coffee and normally will not affect sleep as much.
- Remember that individual differences are large – so be guided by your own personal responses.

It's at the mid-point in the evening (about 8 p.m.) that the 'immovable object' (the homeostatic sleep drive) meets the 'irresistible force' (the circadian signal). This is when there is an upturn in the alertness signal – just as the sleep drive is trying to get us to wind down. The cycling behaviour of the circadian signal is 'endogenous' or internal – that is, it is driven by the genetically controlled activity in the cells of our brain's 'clock'. Generally, it gets stronger as the day goes on, peaking at this mid-evening point, sometimes called the 'Impossible Hour' when most of us simply can't sleep. That may not be too much of a worry for most of us, unless we like to go to bed very early; but when you factor in artificial electric light, which suppresses the 'sleep hormone' melatonin, the impossible hour is shifted later, so that we feel sleepy later and fall asleep much later, shortening our sleep time.

What happens as we start to sleep? Once the impossible hour is past, there is a downturn in the alert signal and we feel increasingly drowsy. This feeling of lassitude is reinforced by the homeostatic sleep drive signal, which continues to rise. Again, this is an endogenous, internal mechanism. This drive remains strong for the next few hours, ensuring that we are driven into sleep, unrestrained by the alert signal, which by this stage will be very low.

However, after about four hours of uninterrupted sleep we become at risk of waking. By this point the sleep drive has diminished and the simple absence of the alert signal isn't enough to keep us 'down'. It is now that we are 'saved' by the circadian rhythm, our internal clock, which sends sleep promotion signals to parts of the brain that suppress wakefulness. Like arousal signals, these sleep signals are sent from the areas of the brain that control sleep (the brain stem and hypothalamus) to our cerebral cortex (frontal lobes) largely in the form of neurotransmitters, powerful chemicals produced within brain cells.

If you wish to optimize your mental performance, have better emotional control and handle social pressures better, it helps to synchronize your sleep schedule with your internal clock. Stressors such as jet-lag and working night shifts 'desynchronize' sleep–wake patterns from the internal clock's circadian rhythm. This results in an alert signal that is too low when we want to be awake and too high when we want to sleep. Our personal experience of these dislocating effects is underpinned by many hundreds, if not thousands, of sleep research publications. Jet-lag alone has been shown to increase fatigue, headache, irritability, poor concentration, motivation losses, and difficulties in learning and memory.

This seems a good point to pass on the tale related to me by Chuck Czeisler, a renowned professor of sleep studies at Harvard and adviser to NASA. He and some colleagues had sought out a commune of indigenous peoples living in the Brazilian forest, about five hours out of Curitiba. The villagers went to bed at night when it got dark, and woke in the morning as it became light. The whole family, up to nine at a time, slept on the dirt floor in the same room. Chuck asked them: 'How on earth do you manage when the kids get up in the night?' After a minute or two of struggling to comprehend the interpreter's question, the mother's answer was: 'They never do.' They had no electricity, no artificial light, no social media, no caffeine or alcohol and no alarm clocks. This is a salutary message for those of us in 'advanced' societies where we have all of these. And not enough sleep.

Not getting enough?

Appallingly, it is almost impossible to receive anything other than pharmacological treatments for poor sleep. But new research has shown us that it is almost entirely possible to avoid sleeplessness in the first place.

As noted earlier in this chapter, widespread sleep disorders first started to emerge some fifty years ago. Their appearance coincided with big changes in society brought on by so-called advances in technology – increasing the spread of artificial (electric) light and information technologies (electrical and electronic devices and social media) – along with heavy use of caffeine and other stimulants. Today, there are ten times as many people getting less than six hours' sleep a night than there were fifty years ago. On average, everyone, regardless of age, is getting two hours of sleep a night less than in 1970. Up to 40 per cent of all adults in Britain say they do not get enough sleep – the highest proportion in Europe. No wonder our economic productivity is low. In millennials and teenagers, the figures are shocking – they spend an average of about eight and a half hours per day on social media or texting, which is more than many of them spend asleep.

Here we must distinguish between *sleep deprivation* and *sleep disorders*. Sleep deprivation is generally thought of as episodic sleeplessness brought about either by external factors, such as shifting time zones, caffeine, alcohol, long work hours, late social activity or social media activity, or by internal or organic factors such as ill-health, anxiety or stress. It may therefore be transient. Sleep disorders are different, though they do of course result in a similar debilitating lack of sleep. These are persistent, chronic, diagnosed conditions of problematic sleep – and there are a lot of them: the International Classification of Disease (ICD) Register lists eighty-eight categories of sleep disorder. They include insomnia, sleep apnoea (breathing dysfunction), bruxism (teeth grinding), movement disorders (e.g. restless leg syndrome) and parasomnia (e.g. acting out dreams). Insomnia is defined as difficulty falling or staying asleep, early morning awakening and dissatisfaction with the length or quality of sleep, as well as its daytime consequences such as tiredness. It is estimated that as many as 40 per cent of adults in the UK suffer from it. If you have difficulty falling asleep or staying asleep at least three times a week for at least three months, and if lack of sleep affects your daytime functioning or well-being, you may have insomnia disorder and should seek the advice of your doctor.

The results of disordered sleep will be familiar to most of us. You had a bad night. You stayed out late. Or your children kept you awake. Or you couldn't resist the late-night movie. Now all the world is out to get you. Why is Starbucks so slow today? Why is the traffic so bad? Why does it have to be raining? Or could all this just be you? Research has shown that sleep deprivation delivers a double whammy. Not only are we more irritable, hostile, tetchy, mean, even angry (more negative); we are also less friendly, cheerful and optimistic (less positive). Sleep deprivation has severed the connection between your cortex – which exercises restraint – and the amygdala, deep in the brain, where fear, hatefulness and anger arise. This 'disinhibition' is compounded by interference with mood-changing neurotransmitters, like serotonin and dopamine, the 'happy hormones'. It's no wonder you are having a bad day.

And you're not only feeling mean. You are also way, way off your game. You can't pay attention. You're not processing properly. Your decision-making is hopeless. You can't tell one thing from another. You've forgotten what you did yesterday. You're not thinking straight. What's going on? New research has shown that sleep-deprived neurons become 'fuzzy' – and, what's more, they fail to talk to each other. Our ability to recall, consider and concentrate depends on millions of cells working smoothly *together*. Sleep deprivation silences much of this internal, coordinated conversation. Perversely, some areas of the brain that should be silent are now left turned on, in a state of permanent activation, adding to the mayhem. The frontal cortex, the thinking part of the brain, is hit especially hard by lack of sleep. In one week-long experiment, a first group of participants was given five hours' sleep per night and a second group eight hours. They were then faced with a daily decision: take a guaranteed sum of money or take the risk of receiving a higher sum – or nothing. Over the week, the five-hour group increasingly went for the high-stakes option – and, worse, as they grew riskier, *they became blind to their own behaviour*. Not only does poor sleep wreck our decision-making ability, we can't see when our decisions are bad (military commanders, take note). Worse still, research also suggests that sleep deprivation makes us more likely to cheat. And as if that weren't bad enough, now for the real killer, literally and figuratively. Lose a good night's sleep and your reaction times will be worse than if you were over the drink-driving limit.

Looking into the crystal ball

We know enough now to make some predictions about sleep patterns and brain health. Specifically, we can tell from your sleep regime in middle age what your brain health will be in ten years' time. A new study has found that in middle age disrupted sleep, including insomnia, is linked to cognitive decline a decade or more later. Intriguingly, though, the study also found that sleeping nine or more hours a night was associated with cognitive problems in later life. This doesn't mean we should worry about the occasional 'bad night' or the occasional 'sleep-in' on health grounds. It's not what we do occasionally that matters; it's our long-term, persistent sleep behaviour.

All longitudinal studies report that bad sleep is associated with cognitive decline from middle age onwards. In its 2016 report on sleep, the Global Council on Brain Health noted: 'Numerous studies have demonstrated that loss of sleep impairs attention, memory and executive function, and increases the frequency of cognitive complaints in middle-aged adults.' They went on to say:

> Large studies suggest insomnia and its primary symptom of frag-
> mented sleep can harm brain function. Older adults who have
> fragmented sleep have increased risks of cerebral small vessel
> disease as well as of poor cognitive and emotional functioning.
> Those with fragmented sleep are at greater risk for faster cognitive
> decline and have a higher risk of Alzheimer's disease than older
> adults without fragmented sleep. Insomnia is also a risk factor
> for stroke and the primary risk factor for the development of
> depression.[4]

Equally, in a perverse paradox, those with poor brain health – suffering from depression or anxiety, or having had a stroke – tend to suffer from poor sleep. So it is a complex picture. Research suggests that if we have insomnia we are up to ten times as likely to develop depression as those who sleep well. Depressed individuals find it hard to fall asleep and hard to stay asleep, and feel sleepy in the daytime.

But don't we need less sleep as we get older? This is a commonly held belief, but in fact it's based on an assumption that is grossly misleading. The picture is far more complex. Once we reach adulthood, our *sleep time* stabilizes, but *the nature of our sleep* changes. According to the American Academy of Sleep Medicine, our sleep time needs across the various phases of life, for most of us, look like this:

- infants (4–12 months): 12–16 hours per 24 hours;

- children (1–2 years): 11–14 hours per 24 hours;

- children (3–5 years): 10–13 hours per 24 hours;

- children (6–12 years): 9–12 hours per 24 hours;

- teens (13–18 years): 8–10 hours per night;

- adults (19 years and over): at least 7–9 hours per night.

These are average amounts, and of course many of us will differ in our individual sleep requirements – and many will not meet these targets, for one reason or another. As a sleep professor once told me, 'You have had enough sleep when you are not feeling tired on waking and do not feel sleepy during the day.'

Though it is not uncommon to find scientific papers alleging that *sleep problems* increase with age, again this is not the whole picture. It is normal for sleep to change with age; but deteriorating quality of sleep with advancing age is *not* normal. The structure and duration (constancy) of our sleep do change – in both quality and quantity – as we get older. We should not expect to sleep as soundly at age 50 as we did at age 25; though some 'good' sleepers may do, they would be a minority, in the order of 5–10 per cent. But the need for sleep remains the same. The Global Council on Brain Health upholds the American Academy of Sleep Medicine's advice that most of us, however old we are, should continue to get 7–8 hours of sleep in each 24-hour period, in order to maintain good mental and physical health. That may include napping – but remember, a nap period shouldn't exceed about forty minutes; much longer than that and the sleep drive will suffer. As we get older, we should also expect to

Sleeping soundly

find our sleep becoming more fragmented – that is, we wake or are woken more easily. These interruptions become more common from middle age. We can also expect the proportion of deep sleep we get to decrease between the ages of 30 and 60. So we may have to put more effort into getting the sleep we need, and into maintaining good sleep and lifestyle habits if we are to keep getting the restorative benefits of sleep.

The first thing to say about sleep and the ageing brain is that there are no as yet unequivocal answers – only emerging and controversial findings. A classic example of this state of affairs is the difficulty scientists have in answering what appear to be two relatively straightforward questions: first, why does our sleep change as we get older? And second, do women age differently from men when it comes to sleep?

To take the second question first, there is a paradox: men are more likely than women to experience far greater disruption and impairment in NREM sleep later in life; *but* women are more likely than men to report poor sleep as they get older. We don't know whether the explanation for this difference in reporting behaviour lies in gender bias or a physiological mechanism. Poor-quality sleep in women may well be related to the menopause, and many female readers will probably recognize this suggestion. The Global Council on Brain Health concludes that the changes in hormones accompanying a woman's transition through perimenopause and menopause can cause insomnia and sleep disturbances. Hot flashes, or surges of adrenaline waking your brain from sleep, may produce copious sweat and changes of temperature, disrupting both sleep and comfort levels. But there is good news. In March 2019, the *Wall Street Journal* ran an article highlighting the discovery that women's brains age more slowly than men's, citing research findings that women generally age more slowly than men and live longer, notwithstanding the sleep-related and other problems of the menopause.[5]

What we do know is that as we grow older, key neurophysiological and neurochemical changes occur in the arousal system in the brain stem, the hypothalamus and the thalamus, as well in the several dispersed areas of the cortex which regulate sleep. For example, age-related loss of neurons in the pre-frontal cortex is linked to reductions in slow-wave sleep; losses of galanin- and orexin-producing cells in the hypothalamus are associated with taking longer to fall asleep, shorter duration of sleep, more shifts to lighter sleep and greater fragmentation of sleep. Finally, we know that objective and subjective measures of poor sleep are both related to higher levels in the brain of beta amyloid – a protein characteristic of the neurodegeneration in Alzheimer's disease.

Now back to the first question posed above: why does our sleep change as we get older? There is now an intriguing new theory: as we get older, our need for sleep remains high, but our ability to generate sleep is impaired. In other words, as the years pass, we need more sleep than we are normally able to obtain. One strong supporting piece of evidence for this hypothesis centres on the sleep-drive molecule adenosine. It has been found that extracellular adenosine is higher in older age groups – among whom lower sleep drive is common – than in younger ones. How on earth do we explain this apparent contradiction? Apparently, there is widespread loss of adenosine *receptors* in the brain cells – and so the higher adenosine levels do not make us feel sleepier, because the adenosine in the blood plasma cannot get into the target cells. This is somewhat similar to what happens in type 2 diabetes, where there are normal levels of insulin but the cells of the body become less sensitive to it and no longer respond so well.

As well as heralding a changing pattern of sleep, increasing age is accompanied by a higher prevalence of sleep disorders. But we should note here that this is the case for a great many other health conditions. Age is the commonest cause of illness, and often of co-morbidity (two or more simultaneous illnesses). As we have previously noted, in the UK the average number of chronic, long-term illnesses at age 65 is one, at 75 it is three, and for those aged 80 or over it is between five and six. So it shouldn't come as any surprise that sleep disorders are also more prevalent after 65. The secret is to slow down the rate at which we age and to reduce inflammation in the body. This will reduce the risk of a whole range of age-related illnesses, including sleep disorders and poor brain health.

Sleeplessness is highly inflammatory to the brain. By sleeping well, we will not only reduce inflammation and the concomitant risk of brain ill-health; we will at the same time improve intellectual performance, emotional balance and well-being, and our capacity for productive social activity. We should not, however, become obsessed with the influence of age alone. As previously mentioned, one of the most noticeable features of our ageing population is that the differences between people become much greater, the older they are. The heterogeneity of later life tells us that many other factors are at play, interacting with the ageing process to produce the differences we can observe; and this in turn strongly suggests that age *per se* is not the sole determinant of sleep disruption in later life. The key message is a very positive one: we can largely control these interacting factors from our earliest adult days so that, working with the brain's amazing power of adaptability, we have a very reasonable prospect of getting optimum amounts of good sleep as we get older.

Sleep well, stay sharp

In summary, many high-quality studies have demonstrated that loss of sleep impairs attention, memory and executive brain function, and causes an increase in the frequency of cognitive complaints from middle age onwards. However, while many scientists have suggested that maintaining good-quality sleep benefits brain health, and some believe that improving sleep can delay or reverse brain ageing, there is still much research to be done. It remains to be established whether better sleep improves brain function or whether better brain function improves sleep – or both. Interestingly, evidence from sleep deprivation experiments indicates that fewer hours of sleep and fragmented sleep appear to affect the mental performance of young more than middle-aged and older adults. So if you are middle-aged or older, be reassured: your younger colleagues get it far worse than you!

Having said all that, and acknowledging the mixed results of sleep deprivation research to date, *it is justifiable to conclude that maintaining good sleep quality throughout your lifespan can promote better brain function and keep your brain sharp.* All the evidence shows that the sooner we start looking after our sleep, the better our brain health will be. This is the common mantra of all authoritative sleep organizations, including the UK Sleep Council, the European Sleep Research Society and, in the USA, the National Sleep Foundation, the American Sleep Association, the American Academy of Sleep Medicine and the American Alliance for Healthy Sleep.

It is easy to get oversensitive and even neurotic about our sleep patterns. One of the sleep experts at the Global Council said to me: 'Don't worry too much about an occasional bad night of sleep. And in any case, you normally sleep better than you estimate.' As a 'health worrier' myself, I found these words very reassuring. I have concluded – and would emphasize once again to every reader of this book – that it is the overall pattern of our lives, over many years, that really matters.

The feel-good factor

THE FINNISH ARE THE happiest people in the world. That may come as a surprise to you, but this is no guessing game. Science tells us so. For the last eight years, the *World Happiness Report* has been ranking the countries of the world according to the key factors which 'happiness scientists' (yes, they exist) have determined make us happy – including income, healthy life expectancy, social support, freedom, trust and generosity. For the curious, the top twenty countries are listed in box 10.1 – from which you can see that the UK, the USA and other big players in the wealth stakes are way behind the leaders of Finland, Denmark and Switzerland.

BOX 10.1: THE WORLD'S HAPPIEST COUNTRIES

1. Finland	11. Canada
2. Denmark	12. Australia
3. Switzerland	13. United Kingdom
4. Iceland	14. Israel
5. Norway	15. Costa Rica
6. The Netherlands	16. Ireland
7. Sweden	17. Germany
8. New Zealand	18. USA
9. Austria	19. Czech Republic
10. Luxembourg	20. Belgium

Source: World Happiness Report, 2020.

It may also surprise you to learn where credit is due for this new emphasis on national happiness. In 1979 the King of Bhutan, the 4th Druk Gyalpo, was returning home from Cuba via Delhi International Airport. During an interview with Indian journalists here, he allegedly stated: 'Gross National Happiness, not Gross Domestic Product, should be the nation's yardstick for measuring progress.' Thus the apparently nebulous concept of 'happiness' was thrown into the serious world of global governance, from an unlikely source. But it's not such an unlikely idea.

Hardly anyone, if we are honest, does not want to know the secret of happiness. It is a natural and universal yearning throughout humanity. For millennia, the great thinkers have pondered the eternal question: 'What is the good life?' – the kind of life that each of us may dream of living. In the American Enlightenment, the 'pursuit of happiness' was ranked equal to both liberty and life itself.

Can science tell us this secret? First, an unsurprising answer. We know *what happiness is* – but *what makes us happy* differs from one person to another. It is an individual thing. There is no universal answer. To be precise, happiness is a state of mind where we are content and satisfied with the state of our lives. If we had been alive in Palaeolithic times, our daily priorities would have been very different from what they are today. Happiness then consisted largely in having got through another day and still being alive at the end of it. But the last ten thousand years have changed all that. One after another, dramatic revolutions – agricultural, industrial, digital – have conspired to make getting what we want that much easier. Now, for the first time ever in human history, whole populations largely have what they want, in physical terms. But the question remains: has this state of material well-being made us any happier?

Though happiness may seem a simple enough idea, in reality it is anything but. In just the past twenty years, over seventeen thousand scientific papers have been published in the attempt to refine our ideas on what makes us happy. The concept hinges on how we as humans perceive what is happening to us. On the one hand, *functioning well* makes a huge contribution to our happiness. Is my marriage OK? Is work going well? Am I earning enough to live on? Are my children well? Scientists call this 'global life satisfaction'. On the other hand, *feeling good* is equally important. Feeling good is a balance between positive emotions – such as joy, elation, affection, appreciation – and negative emotions, such as guilt, anger, fear, rancour. It is all about maximizing pleasure and avoiding pain. And, cutting across both of these areas, there is another fundamental question: am I fulfilling my potential and living up to my ambitions and expectations? For Winston Churchill, it was achieving national fame by the age of 25. For Isaac Newton, it was becoming a learned scholar, which looked unlikely as he ploughed the land on his mother's poverty-stricken farm. And for Thomas Edison it was all about becoming an inventor, when his teachers had told him he was too stupid to learn anything.

Functioning well

Let's now look at what science has found about functioning well – starting with money: can it buy happiness? Unsurprisingly, research shows that money makes a big difference to happiness – *up to a point*. A 2010 study by Princeton researchers Daniel Kahneman and Angus Deaton found that people tend to feel happier the more money they make, up to a limit. That limit was $75,000 (£50,000), after which happiness levelled off. More recent research (published in 2018) found that in spite of inflation, $75,000 or £50,000 still seemed to be the happiness ceiling. This finding may well be difficult to understand (particularly if you don't have much money), since most people assume that more money will solve more problems. But it seems that no matter how much money we have, the fundamental questions of life, such as what it means, who we are going to spend it with and who we are, don't go away.

There's a subtle twist to this tale. Happiness researchers will tell you that if you have money and you're not happy, you aren't spending it right. Harvard Business School carried out a fascinating experiment in a shopping mall in Vancouver. Shoppers were given an envelope containing either $5 or $20 and told to spend it by the end of the day. Half of them were told to spend it however they wished; the other half, however, were told not to spend it on themselves – they gave it to charity or to the homeless or spent it on someone else. By the next day, the results were in. It didn't matter whether it was $5 or $20, those who hadn't spent it on themselves were measurably happier and felt better about themselves. Studies have also shown that spending on experiences makes us happier than spending on material goods – one fundamental reason being that experiences make us what we are, material things do not. I could go on: market research has come up with a myriad ways in which we can improve our happiness by directing our spending. The key point, though, is that money does contribute to making us happy; but the picture is much more complicated than a simple equation of more money = more happiness.

Second, is it true that marriage will make us happy? All studies around the world for the past twenty years have shown a stronger association, however it's measured, between happiness and being married than between happiness and any other disposition, be it cohabiting, divorced, separated or single. This difference is stable across the lifespan and seems not to vary across generations. There may be an element of 'reverse causation' here – people who are happier are the ones who tend to get married, typically to other people who are also happy. But again, the picture is not as simple as it might seem on first glance: as a rule (of course there are exceptions), happiness declines over the course of a marriage; and paradoxically, research shows that the next most happy experience to getting married is getting divorced. We can combine these two research findings in the principle 'a good marriage makes you happy' – tempered by the acknowledgement that the moment it starts *not to be good*, the reverse applies.

Third question: do children make us happy? Parents in all countries and cultures are quick to laud the benefits of having children: 'a child is a bundle of joy'. But the objective facts do not support this throwaway line of folk wisdom. All studies for the past fifty years have shown the opposite: *generally*, parents are less happy than childless adults. But it's a close call. Children have only a small negative effect on the happiness of their parents, so there will be plenty of us who find ourselves in the large minority of those parents who are happy. It does appear to depend on where you live (in Norway and Hungary, parents are happier than elsewhere), on how old you are (older parents tend to be happier) and on your child-care arrangements. The reasons for 'unhappiness' seem to revolve around the conflict between family life and work pressures. Where this is lessened by more enlightened employment practices, for example giving adequate maternity/paternity leave, the unhappiness level is much reduced.

These are just three examples of how well we might be functioning. To get a wider perspective, we need look no further than a study carried out by Harvard psychologists in 2010. Using an iPhone app called 'track-yourhappiness', they contacted two thousand people around the world and asked them to note what they were doing and rate, on a scale of 1 to 100, their subjective well-being (100 being 'feeling very good'). First the obvious: those who were having sex scored the highest at 90. That was way ahead of exercise, in second place with a rating of 75. After that came a cluster of activities: talking to others, listening to music, taking a walk, eating, praying, meditating, cooking, shopping, taking care of one's children and reading. And at the bottom of the list? The three most irksome scourges of our lives appear to be personal grooming, commuting and working.

It would be wrong to finish this section on functioning well without mentioning two other big findings about its role in our happiness. The first is this: even though we may dislike our daily jobs, *doing nothing* is a bad idea if we wish to be happy. Science has found that almost universally, in every culture, total idleness is the way of life least likely to make us happy. It seems that living with a purpose is an essential part of human happiness. One might almost say that if the purpose of life is to find happiness (and many, from Aristotle to the Dalai Lama, say that it is), then that involves living with a purpose. As long ago as 1926, the Austrian neurologist and psychiatrist Viktor E. Frankl was the first scientist to identify *meaning* as an innate drive of all humans – and his theories were tested in the harshest way in Auschwitz, which he survived. Forced to dig ditches in the knowledge that the Nazis had killed his beloved wife, Frankl reflected: 'I grasped the meaning of the greatest secret that human poetry and human thought and belief have to impart: the salvation of man is through love and in love.'[1] The contribution of 'purpose' to our happiness is now embedded in psychological understanding.

The second finding is that whatever we are doing, we should not let our minds wander. A lack of focus will ruin that sense of happiness that comes from functioning well. Back to the Harvard iPhone study, where the researchers found that for 47 per cent of the time on average, across all activities, people's minds wandered. The happier they were in the activity, the less the mind wandered (only 10 per cent of the time while having sex), but the rule held nonetheless. And there's more: whatever people were doing, and however they rated it on the 'well-being' scale – whether it was having sex or reading or grooming – they tended to be happier if they focused on what they were doing. But which way around was this working? Did their level of unhappiness cause the mind to wander, or did the wandering mind cause them to be unhappy? What the researchers found was fascinating. If someone's mind wandered at 9 a.m., then fifteen minutes later they were less happy than they had been at 9.00. But if people were miserable at 9 a.m., they were not more likely to be fretting or daydreaming fifteen minutes later. In other words, there was evidence that a wandering mind causes unhappiness, but no evidence that being unhappy causes our minds to wander. We should clearly keep our minds on what we are doing.

Feeling good

Feeling good is what psychologists call 'subjective well-being'. In the 1990s, Daniel Kahneman, an emerging Israeli psychologist then working in Berkeley, was drawn into what is known as 'hedonistic psychology' – to you and me, the study of what makes our life experiences pleasant or unpleasant. His milestone paper of 1998, written with David Schkade and entitled 'Does living in California make people happy?', was a prelude to his being awarded a Nobel Memorial Prize in Economic Science. In a study designed to investigate the idea that our brains remember imperfectly and are subject to bias, Kahneman and his colleagues recorded the unpleasantness of a colonoscopy examination (a probe up the rectum) to see how patients experienced and remembered it. They arranged for a researcher to sit next to each patient and record their impressions, minute by minute, during the process. The results for two patients are shown in figure 10.1. Clearly, Patient B, whose examination took longer, had more pain than Patient A, whose examination ended after about eight minutes. But when the patients were asked how much pain they thought they had suffered, astonishingly, Patient B's estimate was lower than Patient A's. The reason? Patient B's worst experience was not at the end – *his experience ended well*. Patient B's evaluation was determined by the bias of his memory, for which the duration of the examination did not matter.

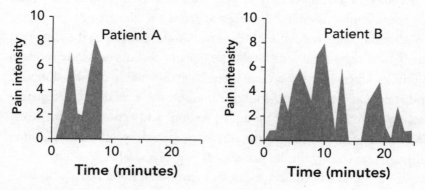

Figure 10.1: Kahneman's experiment on memory

These findings led Kahneman to the conclusion that, as far as feeling good is concerned, the brain has two systems. Our 'experiencing self' perceives how we are feeling in the here and now – paraphrased as 'being happy *in* our life'. This is how we experience happiness moment by moment, in a series of fleeting sensations. The 'psychological present' is said to be about three seconds long – which amounts to some six hundred thousand 'experiences' per month. And what happens to these experiences? Sadly, most are lost. The second system of the brain, the 'remembering self', puts together a 'story' from these experiences which becomes a settled memory; and this story is based on changes, significant moments and endings. The 'remembering self' delivers an entirely different evaluation of our happiness. It tells us how happy we are *with* our life – how satisfied or pleased we are when we reflect on things. At this point we may answer the question posed earlier by Kahneman and Schkade: 'Does living in California make people happy?' When you ask non-Californians about living in California, they compare the idyllic climate, beaches and landscapes to those where they live and (mainly) become dissatisfied. However, when they move to California, their 'experiencing self' invariably is not so happy – except when you ask them to think about their previous life. Then their 'remembering self' invariably tells them they are much more satisfied with their new life.

At the dawn of modern science, Charles Darwin contemplated the origins of our emotions in his great work on *The Expression of the Emotions in Man and Animals*. So revolutionary was it at the time that many now regard it as the book that began psychology. Many modern disciplines, neuroscience among them, have shown that Darwin was right in his conclusion that our individual emotions have discrete, separate identities which drive our thoughts, feelings and actions. In this work, Darwin argued that all humans and many animal species show emotion through remarkably similar behaviours, notably facial expressions, which we use as signals to communicate our emotional state. Darwin was not to know this, but the explanation for his great insight lies in structures of our brain which we share with other species by virtue of our common evolutionary history. Emotions arise from the depths of our primitive brain.

Emotions are automatic and innate physiological responses which have a survival value. Examples are hunger, thirst, lust, fear, anger, disgust, sadness and elation. They are powerful and primitive and visceral. They are the basis of our feelings, and profoundly influence our thoughts and behaviour. Indeed, raw emotion drives our unthinking behaviour. In the twenty-first century, a Quebec television romance parodied the immortal seventeenth-century words of the great French polymath Blaise Pascal, *'Le cœur a ses raisons que la raison ne connaît point'* – the heart has its reasons, of which reason knows nothing. And it is truly a two-way street. Our perceptions of people and circumstances around us can drive our emotions and feelings – you have only to think of the proverbial red mist, the eternal driver of *le crime passionnel*. And when we see something which our thinking brain, the frontal cortex, tells us is a threat, it will respond with fear and aversion, prompting the limbic system to produce what physiologists call the 'fight or flight' response – a complex survival behaviour, first described by the German physiologist Hans Selye in 1936, which is designed to eliminate the threat, one way or another. We either kill it or run away from it. We feel our heart racing, our body (muscles) becomes tense, our breathing accelerates and a 'cold sweat' appears on the palms of our hands. Our body pumps out hormones, notably adrenaline, noradrenaline and eventually cortisol, the stress hormone. We are preparing to defend ourselves or to escape. Another example is food: the sight and smell drives up the hunger within us. And it is the same with lust, a powerful drive fuelling the imperative to reproduce. All these basic emotions are said to be 'appetitive': that is, they drive behaviour until we are satisfied, at which point the emotion, and the drive it generates, subside.

Where do these basic emotions come from? In 2003, nine German scientists performed a literally terrifying experiment on ten carefully selected adults, all of whom had arachnophobia. We're not talking here about a mild attack of the heebie-jeebies when you see a spider in the bath. We are talking about a full-blown medical diagnosis of total, unremitting, paralysing fear of spiders. Each in turn was inserted into a dimly

*Eight physiognomies
of human passions*

lit, inescapable, cylindrical brain scanner – and (you guessed it) – shown images of spiders. In each case, one area deep in the brain lit up like a Christmas tree. The researchers had found the seat of fear in the brain. It was the amygdala, which, as we saw in chapter 1, is part of the limbic system, that collection of ancient structures sometimes called the 'reptilian brain'. These scattered but highly coordinated areas sit between the higher, 'thinking' forebrain and the lower, 'vegetative' brain stem which regulates life-preserving processes such as breathing and body temperature. It is here that all our emotions originate. These are powerful, primitive urges, evolved to ensure our survival; and so they have to be kept in check, which is what the frontal lobes do. Knock them out with a heavy dose of beer and you will see unchecked emotionally driven behaviour let loose.

It has to be said here that 'primitive' does not mean 'simple'. The limbic system is a vastly complex processing plant. Working with the cortex (frontal lobes), it regulates as well as generates our emotions, in the process sifting through enormous volumes of incoming information from both within and outside the body, as well as playing a major role in learning and memory.

A further element of complexity is introduced by the powerful chemicals called hormones, which have a vast influence on how we feel. The 'big four' that 'mess with' our brain are norepinephrine, dopamine, serotonin and GABA – three of which we've already met in earlier chapters. These four hormones have similar chemical structures but underwrite very different emotions. Norepinephrine generates fear and anger, largely in the amygdala, emotions that trigger 'fight or flight' behaviour. Joy or elation is underwritten by dopamine (the 'reward hormone'). Serotonin has many functions (including sleep, digestion and healing), but in the brain it is a mood stabilizer. If our levels of serotonin are normal, then we feel happier, calmer and less anxious, and our emotions become less volatile. The pre-eminent role of GABA is to quieten down the activity of nerve cells, increasing relaxation, reducing stress and calming our mood, even improving our sleep. How do we know if our hormone levels are out of balance? The gold standard – in fact, the only accurate method – is to have a blood analysis, or what the Americans call 'blood work'. This is not, however, a routine procedure, and is usually only available if your doctor prescribes it.

Where do these hormones act within the brain? A husband-and-wife team at the University of Iowa have between them one of the world's most unusual archives – a collection of images of 2,500 human heads. This 'brain library' has enabled Professors Antonio and Hanna Damasio to throw light on exactly where we feel pleasure and pain, with their discovery of one cardinal rule – the brain has separate circuits for 'happiness' and 'unhappiness'. Different parts of the brain produce different emotions and feelings, so the absence of pain or sadness doesn't necessarily mean that we will be feeling happy. Indeed, the Damasios' research has shown what we all suspected – that we can feel multiple competing emotions simultaneously: the pleasurable thrill of fear in a movie; love and hatred of ourselves and others; anger and endearment as the puppy eats your purse. Not only can we feel multiple conflicting emotions, but the Damasio findings tell us that the *two halves of the brain are behaving differently*. This should not be a surprise. It's fairly common knowledge that the right brain controls the body's left side and vice versa. It now seems that the left brain deals with *positive* emotions (anticipation, elation, excitement, pleasure) and the right with *negative* ones (disgust, fear, anxiety, panic). But some halves are more equal than others. Science has revealed that the right hemisphere is dominant – it exerts more control over emotions generally, and it dominates our own facial expressions and our perception of others' (a big part of emotional control). But no rule is without its exceptions. Processing social emotions, for example liking or disliking someone, is a left-brain activity.

The big four 'master' hormones are not, however, the only ones involved in how we feel. There are many others that contribute towards 'feeling good', including oestrogen, prolactin and testosterone – all so-called sex hormones (though they do much more); oxytocin, the 'love' hormone released during emotional and physical intimacy and sex; and adrenaline, which energizes physical action. And then there is the brain's own drug regime – its own internal supply of opiates.

In 1972, scientists at Johns Hopkins University made a startling discovery. They found that neurons in the human brain had receptors for opiates – opium, heroin, codeine and morphine. How odd! Why would our brains be wired up to deal with these powerfully addictive substances? We didn't evolve over millions of years shooting up heroin or snorting 'coke'. What explanation could there possibly be? They soon found out. We weren't designed for addiction. We were designed for pleasure. The brain produces its own morphine-like substances, called *endorphins*. These are our 'happy' molecules. But they're not just around to give us a 'high'. Endorphins promote survival-related changes in our brains and bodies. They are 'feel-good' chemicals produced in the brain to help us get through pain or stress, or to reward survival-promoting behaviour such as exercise or sex. They are euphoric. They are pleasure-producing and pain-reducing – natural 'analgesics' giving pain relief in the direst of circumstances, such as catastrophic injury or trauma or childbirth. Unfortunately, the same life-giving substance has a life-threatening and addictive *Doppelgänger* in the natural world, and our brains are wide open to it. This is natural morphine, cultured in the poppy plains of the Middle East, the cradle of civilization, over five thousand years ago. Michael J. Brownstein paid tribute to the antiquity of cultivated opiates in his classic paper of nearly thirty years back, quoting from Homer's *Odyssey*: 'Presently she cast a drug into the wine of which they drank to lull all pain and anger and bring forgetfulness of every sorrow' – an experience shared desperately by millions before and since.

Alongside endorphins is another set of helper molecules which can also flood the brain. These are the enkephalins, which act in the spinal cord to moderate pain signals travelling to the brain and in the brain itself to regulate our emotional behaviour – for example, our responses to fear, anxiety and stress. Between them, the endorphins and enkephalins light up our lives. Without them, our existence would be grey and robotic. They promote pleasure and diminish pain. They are the ultimate stress-busters. And they enhance our normal activities of daily life, adding colour and vibrancy. It is no wonder that we seek pleasure with such avarice. We are all hard-wired for it.

The laws of physics tell us that if there is a force, there is also a counterforce. In the case of endorphins, it is their dark, brooding cousins, the dynorphins – the opiates of discomfort. The dynorphins are found widely distributed in our nervous system and play an important role in enabling many of our drives and emotions. As in life, so in the brain: there are necessary evils driving us. Discomfort, disgust, repugnance. All vital to our well-being. Sometimes we need to be told that we're under threat – and it is the dynorphins that generate the warning sensations. Hunger when there is too little food. Thirst when we're dehydrated. Loneliness when we're away from our fellow humans. Disgust and repugnance when we're confronted with noxious substances. Where would we be without dynorphins? By making us feel bad, they prompt us to seek to avoid what we don't need or shouldn't have and get what we do need. This family of seemingly nefarious opiates is actually our best friend. By generating unpleasantness, they steer us away from danger until a state of well-being is restored.

I feel, therefore I am?

One of the most admired characters in the television series of *Star Trek* was the humourless Mr Spock. Half human, half Vulcan, Spock would ride roughshod over the mere humans around him. Super-rational, super-cool and remorselessly calculating, he would tower over the hapless crew of the *Enterprise* with their flawed, emotional brains. Except – if he was half as clever as he claimed he was, Spock would have invented emotions. For if we had evolved without emotions we would never have truly evolved. They are key to our complex behaviour. Our body's visceral emotions rise into the conscious reality of our brains – our frontal cortex – and they become *feelings*. These feelings intertwine with values, judgements and beliefs; they allow us to manipulate and co-opt the behaviour of others. They are amusement, anger, apathy and awe. They are depression, desire and despair. They are panic, passion and pride. And these are just a few of the scores of feelings catalogued by scientists. Our complex human behaviour is driven as much by our feelings as by logic and rational thinking, in spite of our boast of human intellectual superiority, our arrogant claims to be the Spocks of our planet. And that is why we humans are a truly formidable species. Within our brains is a deadly machine, in which there is an orchestrated and reciprocal interplay of emotions, feelings and rational thought. Thinking helps our emotions and our emotions help our thinking. Together, these disparate elements give us a powerhouse of decision-making, intuition and action. What separates us from all other species is not only our ability to regulate and control our feelings, but our brain's capacity to use them to decide, plan and execute.

All of which is not, of course, to say that other species have nothing to teach us. In 1954, two Canadian scientists conducted an experiment on rats to investigate the 'reinforcing' (rewarding) effect of a mild electrical signal to the brain. They placed an electrode in contact with the brain of rats that could self-administer the shot to the brain by pressing a small lever. The rats were unrestrained and had access to every creature comfort, including food and water. Rat No. 32 pressed the lever 7,500 times in twelve hours, with an hourly average of 625 shots – about twelve times a minute or once every five seconds. This Olympic performance was however totally eclipsed by another rat which delivered a whopping 1,920 presses in a single hour – that is, one response every two seconds – and died of exhaustion. Where, you may ask, was the brain being stimulated? The experimenters had found *the pleasure centre of the brain*. But they also found a sobering corollary. There were some areas of the brain where the effect was just the opposite: the rats did everything possible to avoid stimulation. Thus the humble rat teaches us the cardinal principle of feeling good: maximize good experiences and avoid bad ones.

How our emotions empower our brains

At this point you may be asking: what do our individual feelings of happiness and well-being have to do with brain health and our brain power? Surely, good mental well-being is born from a healthy, well-functioning brain? Science is now telling us that a lack of mental well-being (reflected, for example, in feelings of pessimism, uselessness, misery) has been shown to interfere with our three primary brain attributes: our abilities to think and reason, to interact with others and to control our emotions. We can all take steps to improve our mental well-being – and it's worth doing, because it's not just a question of feeling good; it's your secret weapon for empowering your brain. What we know about well-being is summarized in box 10.2.

BOX 10.2: WHAT WE KNOW ABOUT WELL-BEING

- Our happiness or well-being is a state of mind where we are content and satisfied with our lives.
- Our well-being has two elements to it – *functioning well* and *feeling good*.

- Functioning well is all about our global life satisfaction – am I doing well in my work, home life and personal affairs? Am I coping with life's changing circumstances?

- Feeling good consists of how happy we are in the moment – are we comfortable, healthy, happy, purposeful?

- Feeling good is a balance between positive feelings – such as joy, elation, affection, appreciation – and negative feelings, such as guilt, anger, rancour.

- Emotions are automatic and innate physiological responses which have a survival value. Examples are hunger, thirst, sexual drive, fear, anger, disgust, sadness and elation. They are powerful, primitive and visceral.

- Feelings are conscious internal experiences based on emotions which rise into the thinking part of the brain (our frontal lobes).

- Emotions and feelings profoundly influence our thoughts and behaviour and are vital for normal cognitive function – thinking and reasoning, decision-making, planning and good social relationships.

- Our brains are hard-wired to seek pleasure and avoid pain. Both feelings are vital for well-being, requiring a balance between these two conflicting emotions, which battle for our attention.

Most of us take pride in our decision-making capacity. We are faced with a choice or a decision to be made. We look at the facts, think them through and then make a rational decision. Wrong. Most of us 'think' with our emotions and make decisions based on our intuition – what our gut tells us – literally and figuratively! Then we often use the 'facts' to back up what we've already decided. We may use a process or rules to guide us. But in most of our everyday decisions, *our emotions rule supreme*. Psychology failed to recognize this for most of the twentieth century; then, about 1970, research began to question the received wisdom, though it didn't really get going until 2000. The years between 2004 and 2011 saw eight times as many studies in this area as had been published in all previous years. And what did they show? That the goal of our emotionally driven decision-making is to move away from negative feelings (guilt, fear, regret) to positive ones (pride, pleasure, contentment). Almost anything to get away from stress, anxiety, worry and fear. Decisions on who we are with, what we do, what we eat and drink, how much we earn – in terms of respect, money and position – are all driven by a deep evolutionary yearning to *feel good*.

And here's the key point to which all this is leading: once we are feeling good, our conscious mind works better. *A blissful state of well-being is the secret of brain power.* One milestone study amply demonstrates this. In 2008, in a large population study at Cambridge University, researchers measured psychological well-being in more than eleven thousand adults aged over 50. Each participant was also tested on a full battery of cognitive tests, looking at time orientation, memory, verbal fluency, numerical ability, thinking speed and attention. The results were unequivocal. All those in the top 20 per cent of well-being scores were clearly superior in their cognitive skills. To quote the authors, 'psychological well-being was significantly associated with performance in all individual cognitive domains'.[2] This study was a first and, moreover, it was very powerful – on two counts: one, it involved a large number of people; and two, it was longitudinal, showing that the relationship was sustained through people's lives. Later work has reinforced these findings. In 2018, AARP Research carried out a survey of 2,287 American adults aged 18 and older, asking how they perceived their own mental well-being and their brain health. It found that adults age 50 or older who scored higher on the scale of mental well-being tended also to report better memory and thinking skills.

Almost all studies conducted in Western economies show a U-shaped relationship between well-being and age – in other words, in general we feel good in our youthful years, worse in our middle years (40–50) and better again as we age. A typical set of results from a study in the USA is shown in figure 10.2. There are numerous explanations for the shape of the curve – the main one being that economic circumstances, diminishing work pressure, the positive aspects of retirement and the easing of family responsibilities as children become independent all contribute to increasing life satisfaction after 50. And the better we feel, the better it is for our brains.

There is, however, another explanation for our increasing well-being as we get older, and it lies in the way the brain operates – *older adults tend to pay greater attention to and remember more positive information, and at the same time to downplay negative events and experiences.* We have some idea of one of the brain mechanisms underlying this shift: in a study in the USA, those middle-aged and older adults who on viewing 'positive' and 'pleasant' pictures showed sustained activity in an area of the brain called the striatum – usually involved in processing feelings of reward – also reported overall greater levels of well-being.

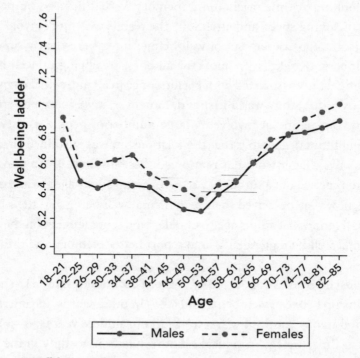

Figure 10.2: Well-being and age

However, we disregard the influence of negative emotions at our peril. In decision-making, certainly, they are there for a purpose. For example, anger motivates us to remedy injustice, fear to avoid a risky option, and (anticipated) regret or guilt to move away from indulgence in lust – all of which common observations prompted the eighteenth-century Scottish philosopher David Hume to declare: 'Reason is, and ought only to be, the slave of the passions, and can never pretend to any other office than to serve and obey them.'[3] But how do these negative feelings affect our thinking?

Let's first look at anxiety as a response to unpleasantness. In 2020 a group of Brazilian neuroscientists published the results of an ingenious experiment using EEGs to measure the electrical signals within the brain in two groups of patients, one of which had been subjected to 'unpleasant images' and the other of which had not. (They would not disclose what the unpleasant images were, other than that they resulted in anxiety and 'intrusive' thoughts.) To make sure the 'unpleasantness' was working, all participants had their anxiety levels measured. Each group then carried out a memory test, the researchers measuring their responses by recording an electrical signal known to be sensitive to the number of task-related items stored in working memory. This signal was significantly lower in the anxious and disturbed group, indicating a much poorer memory performance than that of the non-anxious group. In fact, the higher the measure of anxiety, the lower the working memory capacity. It's worth pointing out that working memory is the most important of our 'higher' brain functions – vital for complex intellectual tasks, any kind of learning and simply paying attention.

But it's not just working memory that suffers when we're anxious. High levels of anxiety drain our brain of resources, limiting the attention we can pay to what we are doing. The more critical the task, the bigger the impact. And the more anxious we are, the more our attention wanders. Severe distraction even prevents us from being aware of where we are going wrong – from monitoring errors – thus disabling our judgement and decision-making ability. Our capacity to refresh and update information as we work at something is also damaged by anxiety. Updating in our minds is critical to fluid intelligence, so in its absence we grossly underperform at reasoning and problem-solving.

The effect of fear is little better. Fear creates alarm and distraction, diverting all our energy to deal with the immediate threat. But it also heightens our awareness of and connections with others. This is called social learning, and teaches us to modify our responses by looking at those around us – who may be frozen in panic or doubled up with laughter! Fear also involves context and control, processes in which our amygdala and hippocampus work together, comparing stored memory with what's going on around us now. The problem arises when control is lost and fear becomes prolonged. In his experiments on prolonged stress, Hans Selye found that adaptation to fear could only go on for so long before the stress systems of the body collapsed. This is a severe problem in Western, developed societies where one in four people now suffer from some form of anxiety disorder in their lives and nearly 8 per cent of us suffer from some form of post-traumatic stress disorder.

In his book *The Marshmallow Test: why self-control is the engine for success*, Walter Mischel creates a persuasive case for keeping our cool. Anger, a 'hot' emotion led by the limbic system, does more than rev up our hostile behaviour. It limits cognitive abilities such as working memory, judgement and evaluation, reasoning, computation, decision-making and logic. It would be difficult to find a more comprehensive list of reasons why we should control our anger. But in addition to this substantial case for chilling out, neuroscience has added a telling observation: when we get angry, we divert the brain's blood supply away from the essential areas for thinking straight. In a remarkable study involving 7,413 patients of an outpatient psychiatric clinic, researchers in the USA identified a 'high' anger group (top 25 per cent) and a 'low' anger group (bottom 25 per cent) on the basis of a reliable self-reported anger scale. A SPECT brain imaging scan was then administered to both groups, examining seventeen brain regions. The results showed that participants with higher self-reported anger had lower blood flow in the left limbic region, basal ganglia, frontal lobe and parietal lobe – several of which are heavily involved in important cortical functions. No other significant blood flow abnormalities were found. Anger is a natural threat-related behaviour; but it seems that once it has served its purpose it should be allowed to ebb.

Improving your brain power: be positive, be active, be social, be purposeful

Mental well-being is achieved when we are feeling good (which includes being able to balance our emotional experiences), are functioning well and have a sense of purpose in life.

We'll focus here on the first of these, feeling good. Acknowledging and accepting both positive and negative emotions is vital to mental well-being. Anxiety, fear, feeling down and being miserable are all part of normal human experience. We have to understand that nature has designed these negative feelings to be helpful – to motivate us to make changes that will move us towards more positive feelings. The danger comes where negative feelings endure beyond their usefulness. Persistent negative moods or feelings are damaging and can result in mental health problems and disorders. When we are no longer able to get rid of these feelings ourselves, we have to seek medical help or psychological therapies. But there are many ways short of visiting the doctor to sustain our feelings of happiness. We'll now consider the evidence for what we can do to achieve a sense of feeling good and thereby empower our brains to work well.

We have already seen, in chapter 6, the benefits of positive thinking. Not only is it good for our general health, it works wonders for our brain power. It pays to accentuate the positive in life – even during times of illness, distress, frustration or financial hardship. To some extent our ability to do this is limited by our personality. Research has found that extraverts are more likely than introverts to rate their well-being as high, and optimists too find positive thinking comes more easily than pessimists. Scientists have worked out many strategies for training us to think positive. They include:

- Finding out how 'positive' you are by taking a well-being quiz, such as the one constructed by the Berkeley Well-being Institute, which is available online and gives you a 'positivity' score.[4] This will give you a basis to work from.

- Improving your 'positive word' vocabulary. Scientists have compiled lists of thousands of words rated for three properties – their pleasure rating, their arousal value and their dominance (how powerful they are). Identify and practise using positive words. Writing down lists or rehearsing them mentally will help the brain to process and use them better. It's interesting that the most pleasurable words arouse us as much as the least pleasurable ones – so it pays to 'up' our positive word list. These lists and other applications are also available online – for example at positivepsychology.com – or via an app store: search for e.g. 'I am – Positive Affirmations' or 'Feelgood – Positive Thoughts'.

- Prolonging the pursuit or the enjoyment of pleasure in order to achieve a much fuller experience of a happy or pleasant moment. Many emotions, such as hunger, thirst and lust, are appetitive and drive us to become satiated, at which point our energy evaporates – along with our experience of pleasure. This oscillation between the pursuit of pleasure and its consummation has been called 'the seesaw of positive feelings' by some authors. Basically, this is about following the age-old advice to savour or just enjoy the moment and not rush pell-mell through enjoyable experiences.

- Not being afraid to show negative emotions when you need to or where the situation demands it, for example in cases of misfortune, bereavement, loss or trauma. Such situations promote the release of dynorphins, which will motivate us to seek relief from the discomfort of these experiences, so we should not casually dismiss them when they arise. However, we should certainly control unnecessary negative emotions such as fear, anxiety or anger. The acid test is to ask if these negative emotions are helping us – and if they are, to use them for so long as they are helpful, but no longer.

- Revisiting anxious or even fearful memories and reforming them to minimize their effect. Revisiting memories to reconstruct them in a new light is in fact something that almost all of us already do on a daily basis without being aware of it. It's a technique used all the time in high-risk professions such as 'free solo' mountaineering, climbing, extreme sports, expedition leadership and even poker to reduce their future anxiety levels. It's also increasingly used as an anxiety-reducing strategy by psychologists.

- Savouring or holding on to good experiences, or recalling your happiest moments. This technique, which is known to improve our feeling of well-being, uses the 'remembering self' to create or bring back happy or pleasurable moments. These memories or extended experiences will help the release of endorphins, which in turn cause the release of dopamine, a powerful 'reward' hormone.

- Using visualization to prepare for big, challenging events, such as an interview or a public speaking engagement. This technique, widely used by Olympic coaches, involves running through the future event moment by moment mentally and visualizing your own *competence*. It's known that feelings of confidence improve performance by reducing negative emotions such as anxiety and fear of failure.

These are just a few of the many ways of fostering positive feelings; we shall look at more of them in the section on 'managing stress' later in the chapter.

For now, we'll turn back to another way of feeling good: being active. As emphasized in chapter 2, engaging in regular exercise benefits not only the body but the brain and mind. If you don't currently exercise, there are countless opportunities to start. Basically, virtually anything which increases your physical activity is going to be good for the brain. It is worth mentioning that exercise not only releases brain growth factors such as BDNF but triggers the feel-good post-exercise experience through the release of endorphins and enkephalins into the brain.

One topic we haven't looked at yet, which relates to being active, is spending time outdoors – in what has been elegantly summarized as the 'Green Gym'. For the townies among us, missing out on nature is a big deal for our brains. There is now a huge volume of evidence to show that enjoyment of the natural world – a capacity deeply embedded in our brains – not only counters stress and anxiety but is beneficial to our brain health, improving self-esteem and combating negative emotions and feelings such as tension, anger and depression. Outdoor 'Green Gym' activities can include hiking, climbing, fishing, horse riding and orienteering but also less sporty pursuits such as gardening, tree planting, sowing meadows and even establishing wildlife ponds.

One unlikely such outdoor activity is shooting. You may be against it. You may be for it. But either way, research has shown that shooting guns is rewarding to the brain and releases waves of oxytocin and dopamine, hormones which elate us. No wonder millions of people worldwide participate in lawful shooting, whether as recreation or sport. And it doesn't have to involve guns. Any target-related activity, such as darts, archery or ten-pin bowling, appears to have the same atavistic feel-good reward, a legacy of our hunter-gatherer history.

Many physical activities are also social, of course, and we've already seen, in chapter 6, the exceptionally strong influence of our social lives on brain health. I will not repeat all that advice here, but will just mention the key finding of one unique and highly relevant research project – the Harvard Study of Adult Development, the world's longest-running study on happiness, which has been running since 1938. At the outset, the vast majority of the participants (724 of them, all men) – regardless of social class background, wealth or poverty – stated that at the start of their adult lives, becoming famous or wealthy was their prime aim in life. At the end of the study – by which time they would all have been in their nineties, some having gone from poverty to wealth and some in the opposite direction – it was not. With experience, the participants had come to realize that *good relationships* are what matters, keeping us happier and healthier. What we tend to want as humans is the quick fix to make our lives happy and robust. This is not nature's way. It takes long, stable relationships to keep that smile on our face – and this is lifelong, hard work, repairing fences, removing grudges, keeping in touch with family and friends and loving those close to us.

And it's not just our relationships we need to work at to maximize our well-being. We considered earlier in this chapter the danger of drifting through life in idleness or aimlessness. History and philosophy are replete with examples of the satisfaction gained through fulfilment of one of our most salient qualities – the need for a sense of purpose. At this point it's worth considering some sound advice from the great physicist Stephen Hawking, which encapsulates much of the advice offered above: 'One, remember to look up at the stars and not down at your feet. Two, never give up work. Work gives you meaning and purpose and life is empty without it. Three, if you are lucky enough to find love, remember it is there and don't throw it away.'[5] For our own well-being, seeking a goal or goals and pursuing them is the essential core of having purpose in life. We won't always succeed, though, and to maintain our well-being it is also vital to come to terms with our past decisions and accept what we cannot change.

The key points on how to improve your well-being and thereby your brain power are summarized in box 10.3.

BOX 10.3: IMPROVING YOUR WELL-BEING – IMPROVING YOUR BRAIN POWER

- To promote well-being, seek to balance your positive and negative feelings.

- A good sense of well-being is associated with superior thinking skills, quickness of mind and good memory.

- Excessive negative emotions (such as anger, fear, anxiety) debilitate our thinking skills, distract us and result in poor cognitive performance.

- Improve your positive thinking by:

    o cultivating the use of a positive vocabulary;

    o prolonging and enjoying the pursuit of rewards and pleasurable experiences;

    o tolerating and accepting negative feelings where they are appropriate, but not letting them persist unnecessarily;

    o revisiting and reconstructing anxious or fearful memories to reduce their impact;

    o savouring and holding on to good experiences or recalling your happiest moments to improve your feeling of well-being;

    o using visualization to improve your confidence for future challenges, such as public speaking events, interviews or tests;

    o managing your stress levels (see next section) and recognizing that they will vary at different stages of your life, generally improving after middle age.

- Stay active and have a programme of regular exercise – anything which improves your daily physical activity levels will improve your well-being. Pursue outdoor activities and experiences wherever possible – whether 'green' or nature-related activities or just being in the countryside or at the coast.

- Don't neglect your social life. Build up and maintain contacts with friends and family. Stay connected and seek group opportunities, like clubs or community activities, wherever possible.

- Develop personal and work-related goals and actively cultivate a sense of purpose. This can be looking after friends or family, having an absorbing and demanding hobby or pastime, or striving for career goals. Attempt to balance the different areas of your life.

Managing stress: yoga, laughter, art, music and more

It may come as a surprise to learn that stress is beneficial to us and that our well-being depends on it. Up to a point. There is a U-shaped relationship between stress, arousal and performance. It's called the 'Yerkes–Dodson law', after the two psychologists who developed it, and we show it below in figure 10.3. What it depicts is that as stress (i.e. some external demand or even threat) increases, so does our arousal level in the brain – and, in turn, so does our performance. The peak of the curve marks the optimal level of stress for optimal performance; but if stress continues to increase past this point, our performance declines. Another way of putting it is that any stress level before the peak is 'good' stress and anything after it is 'bad' stress. The secret, therefore, is to find our own individual 'sweet spot' beyond which we know we are overstressed, and then seek to change our demeanour and our circumstances so that we do not go into the downward, negative slope of the curve. We need to recognize when we are overstressed and beyond our coping level. Exceeding that level will not kill us, but long-term high stress will certainly damage both performance and brain health. Individual differences in acceptable stress levels are huge. Some people are only satisfied if they have climbed a death-defying mountain on their own with no support team. Others crave the pressure of risking everything on a single gamble. Others still find it stressful enough just to get through a single day. Most of us are somewhere in between these extremes. But all of us have to take great care in making sure that we have coping mechanisms to deal with any 'bad' stress. Stress, in a word, must be managed. And nature has provided us with our own internal barometer – our feeling of well-being. We'll now look at the evidence of what works (and what doesn't) in managing stress.

The first British observers of yoga in the subcontinent described it as a dubious activity, pursued by deceptive vagabonds pretending to be pious.[6] Now practised by an estimated half a million people in Britain, it is characterized more positively as an ancient mind and body practice with origins in Indian philosophy.[7] What a turnaround. And not just in Britain: in the USA yoga is now the most popular complementary therapy, with at least 13 million devotees. But however impressive, the numbers of Western followers are totally eclipsed by the 200 million practitioners in India, yoga's birthplace. Yoga differs from conventional exercise in its axiom of uniting body and mind, through deep breathing, careful holding of a variety of physical positions, stretching and quiet meditation. Multiple research studies have reached a single, common incontrovertible conclusion about this unique activity: yoga benefits both mind and body. *There is considerable evidence that yoga is beneficial to well-being, including positive effects on anxiety, stress, depression and*

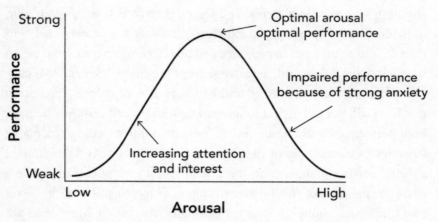

Figure 10.3: Performance, stress and arousal: the Yerkes–Dodson law

overall mental health. Moreover, research has shown *a clear relationship between yoga and brain structure and function.* A systematic review of eleven major neuroscience studies on yoga published in 2019 concluded: 'Collectively, the studies demonstrate a positive effect of yoga practice on the structure and/or function of the hippocampus, amygdala, prefrontal cortex, cingulate cortex and brain networks including the default mode network (DMN).'[8]

The default mode network. Three small words of fundamental importance to our brain power. The DMN is a brain network that deals with 'self-thinking' and 'mind wandering'. If we 'self-reference' (concentrate on ourselves) the DMN goes quiet. If our mind wanders, then the DMN springs into action. The calming aspect of meditation, alone or within yoga, with its emphasis on mindfulness (conscious direction of attention), quietens the DMN and prevents the mind from wandering. Earlier in this chapter, I cited research showing that if we let our minds wander, self-reported happiness declines. It should therefore come as no surprise that this 2019 review paper also showed how yoga and meditation improve cognitive performance, attention and memory – albeit indirectly, by improving the regulation and balance of negative and positive emotions. How does it do this? Via neurotransmitters. Evidence has shown that when meditation activates the brain's frontal cortex, the brain stem releases GABA, a neurotransmitter which reduces anxiety. We feel calm. Meditation also acts via the dopamine reward system. We feel positive. It's no wonder that meditation turns out to be exceptionally good for brain health, cognitive performance and well-being.

One last thought on the subject of yoga. In chapter 3 on the 'Godzilla brain', we mentioned the pervasive influence of the vagus nerve. If you thought that the claimed benefits of the 'Om' chant in yoga are fanciful, ethereal wishful thinking – think again. For there is now a neuro-physiological explanation of how this chant influences our brain. Yes, indeed, the yoga chanting and the rhythmic, repetitive, respiratory motions are more than a ritual – they are a therapy. These carefully choreo-graphed actions and movements – which one scientific report calls the 'breath of life'[9] – stimulate the vagus nerve, feeding the automatic control networks of the nervous system. Well-being and emotional balance ensue. And with them, rich benefits to our brain power. As we meditate, move and hum our way through our yoga session, we are unwittingly nurturing our executive function, working memory, concentration and creativity. That is the power of contemplation. It is anti-stress, anxiety-reducing and brain-boosting.

So we should take yoga seriously. We should also take laughter seriously. It has been key to our survival as a species. It is a fundamental part of being human. Selected by hundreds of thousands of years of evolution, laughter is undeniably in our genes. For without it, social cohesion would be much, much weaker. Have you noticed how contagious it is when someone starts to laugh? We are united. We are a group. We share a common experience. And we are stronger. It also is a primary disarming signal. Conflict is avoided and suspicions lowered by shared laughter. So contagious is laughing, in fact, that medical science has recorded 'laughter epidemics' in cultures around the world where people just can't stop laughing. Now science is revealing the secrets of laughter.

First, we laugh out loud far more from group amusement than from hearing jokes; and we are thirty times more likely to laugh in social situations than when we are alone. Laughter is also spontaneous and can rarely be forced. It's as if our brains instinctively know what is funny and prompt us to react accordingly – and women do more of the reacting than men, it seems: according to the scientific evidence: women are more than twice as likely as men to laugh responsively and men more likely by an equal margin to create the laughter. In the observations of two German scientists, the women who laughed more on a first date reported themselves as being more interested in their man than those who didn't. And as a relationship continues, female laughter is a great barometer of how well it is going. Laughter is an integral part of human courtship – and of feeling good.

But how does all this work? It's complicated; and it happens quickly. Our perception of something funny arises in our frontal lobes. Then an area called the 'supplementary motor area' fires up, tapping memories such as the movements associated with laughter. The next part to be activated is the 'nucleus accumbens' in the limbic system. It assesses the pleasure of the story and the reward that the spark of humour brings. Our heart rate rises, we break into laughter, and our brain releases 'feel-good' neurotransmitters: dopamine, serotonin and an array of endorphins.

For all these reasons – strengthening social bonds, defusing hostility, optimizing attractiveness, fostering brain connections and sparking good feelings – it is no wonder that a daily dose of laughter reduces stress and is a powerful promoter of well-being. We should make the effort to surround ourselves with humour, the precursor to laughing. Buy humorous cartoons and books; spend time with people who make you laugh; watch funny films and stand-up comedy; and try to laugh about the problems of life. All simple things – but all an investment in feeling good.

We'll round off this section with three more feel-good suggestions. If you have watched the film *Alpha*, you will have seen a fictitious and inspiring account of a real event which took place some forty-five thousand years ago, when humans were hunter-gatherers. It was the domestication of canines. Few other biological relationships between different species come as close as that between human and dog. The bonding which can take place has now been 'brain-mapped' – on both sides. In both us and

Man's best friend

the dog, closeness evokes high electrical activity in the reward system of the brain, releasing waves of oxytocin, soon followed by endorphins and **serotonin**. Result: calm, elation, euphoria, peace. It is no wonder that the therapeutic use of dogs has increased massively in the past two decades – and in the USA by 1,000 per cent in just ten years between 2002 and 2012. We can also form close relationships with other pets, too – and this is reflected in the 'emotional support animal' or ESA, which can be anything from a toucan to a python. In the USA, the rights of people dependent on ESAs are now enshrined in law: interfere with them and the federal government may sue you. It's no wonder that there has been a rise in the abuse of these protections – for example by travellers insisting that they be allowed to take a snake or a parrot on to an aircraft! But that doesn't take away from the really serious point: namely, that pet ownership, especially of dogs, is a huge stress-buster and a valuable tool in our well-being kit.

And then there's art. Roy Lichtenstein's huge pop-art installation *WHAAM!* stunned the New York art scene when the giant twin can-vasses were first shown in Leo Castelli's gallery in 1963.

Art critics opined vehemently – favourably and unfavourably – and three years later the Trustees of the Tate Gallery argued over whether they should pay £4,665 for it (that would be £87,538 if we paid it today). But

An early study for WHAAM!

they did; and this undeniably impressive exhibit attracted curious and enthusiastic crowds – a total of 52,000 over just two months in January and February 1968, far more than for any other single work of art then exhibited by the Tate. Visual art compels us to look, to observe, to absorb. And sometimes to be awed and *WHAAMED!* Throughout history, artists have created intense emotional experiences for us. The sense of visceral awe and inspiration which 'draws us' into a painting such as *WHAAM!* is such a deeply felt experience that neuroscientists call it 'embedded cognition'.

Professor Semir Zeki, Chair of Neuroaesthetics at University College London, has spent a long career studying the neuroscience of art. He explains the influence of art on our conscious experience as follows:

> Art can heal us, inspire us, and alter our brain chemistry . . .
> Contemplation, observing, and taking in beauty all stimulate
> pleasure centres within the brain while increasing blood flow by up
> to 10 per cent in the medial orbitofrontal cortex. This can lead to
> an elevated state of consciousness, well-being, and better emotional
> health.[10]

And if that's not enough to convince you, consider that doctors in the UK now prescribe art as preventative and active therapy for stress and to foster well-being, in part on the basis of evidence from a report commissioned by the UK All-Party Parliamentary Group on Arts, Health and Well-being. There are also a wide range of specific programmes of 'art engagement' in place across the country, offering opportunities to join community art groups which provide the opportunity to paint and to visit art exhibitions and museums. Without even leaving your home you can try painting, drawing or just doodling, maybe keeping a daily art diary in which you draw your moods or holding 'art parties'; and you could visit one of the many websites which curate art and artists for therapeutic purposes, such as The Healing Power of Arts and Artists.[11] The expression of creativity in visual art is a fundamental human activity, embedded in the deepest parts of the brain some 64,000 years ago. Each time we draw, paint or merely observe, we activate these ancient pathways to our lasting benefit.

Art, of course, covers more than just the visual; and if you had to identify one indisputably stress-relieving activity, enjoyed by millions around the world, you might well pick the enjoyment of music. Music has been shown by numerous experimental studies to be beneficial to physiological, emotional and cognitive stress responses – and this effect seems to be universal to all cultures. The enjoyment of music inspires us to move and dance, to sing and to enjoy the company of others, bringing intense pleasure and happiness. And the reason for these multiple benefits? No other comparable activity stimulates and coordinates multiple areas of the brain in the way music does. Listening, attending, remembering, moving, speaking, thinking, feeling – all these systems are 'go' when we hear music, and none more so than our emotions. Music can change our mood in a flash. Though more research is needed to

Two country dancers

understand more fully how our brains hear, perceive and process music, and indeed whether music has a direct role in promoting our memory and thinking skills, there are numerous practical ways in which we can draw on the power of music to reduce our stress levels and improve our well-being. They include purposefully using music to evoke positive memories and associations; listening to new music, allowing unfamiliar melodies to stimulate the brain; learning to make music with others, thereby benefiting from the social activity as well; and using music to motivate movement, activity and exercise by dancing.

For centuries, people have drunk alcohol to relieve stress. In *Julius Caesar* (Act IV, scene III), Shakespeare has Brutus exclaiming: 'Speak no more of her. Give me a bowl of wine. In this I bury all unkindness . . .' In fact, our taste for alcohol began not hundreds but hundreds of thousands of years ago, as our human ancestors were drawn down from the trees to eat the fermented fruit fallen on the forest floor. And our liking for alcohol continues undimmed today. Around the world, vast quantities of it are drunk every year, and in almost every culture.

Why is alcohol consumption so prevalent? And unrelenting? Because, beyond doubt, drinking moderately is a highly rewarding experience for most people. Some even seek and enjoy drunkenness, though not its after-effects. We have a few drinks and we feel energized and sociable. Our rising blood alcohol brings with it pleasurable feelings of excitement and elation, as dopamine is released in the brain. Noradrenaline invigorates us and our tension and fatigue evaporate. Another drink or two and we become impulsive and reckless as rising alcohol levels suppress activity in the pre-frontal cortex – the seat of sober decision-making. Energy consumption in the cerebellum – which coordinates movement – declines markedly. In the hippocampus, our blood alcohol starts to impair memory formation. And falling GABA levels make us sluggish, slow and confused – we begin to 'word our slurs'.

But was Brutus right to drown his sorrows in alcohol? Is there any evidence to support this almost universal belief in the stress-relieving properties of booze? For those of us who enjoy a tipple or two there is good news. A paper in 1999 concluded that drinking can indeed reduce stress 'in certain people and under certain circumstances'.[12] And it gets better. The equivalent of about two drinks per day has been shown in animal experiments to reduce brain inflammation and to improve the drainage system of the brain – adding to the growing body of research showing the benefits of moderate alcohol drinking to both brain and general health.

But we have to know when to stop. There is plenty of evidence to show that if we don't, our social and stress-related drinking will lead to a heavy, long-term habit.[13] And for the brain, the consequences of such a habit are severe. Three studies fire a shot across the bows of the heavy drinker:

- Heavy drinking was found *to speed up memory loss* in early old age, at least in men. In a 2014 study, men who had more than two and a half drinks a day experienced cognitive decline up to six years earlier than those who did not drink, had given up drinking, or were light or moderate drinkers.[14]

- A report in the *Archives of Neurology* showed that prolonged heavy drinking *shrinks brain volume*.[15] The study found that people who had more than fourteen drinks per week over a twenty-year period had brains 1.6 per cent smaller (one measure of ageing) than non-drinkers.

- If that's not enough for you, a recent study in the *Lancet* showed that alcohol-related damage to the brain actually shortens life expectancy.[16] Those who drank ten or more drinks per week had their life expectancy shortened by two or three years compared to those who had five drinks or fewer a week. And those who had eighteen or more drinks per week saw their life expectancy shrink by four or five years.

There can therefore be only one line of advice: moderation is everything. What is moderate? Almost universally, authorities say it is one drink per day for women and two drinks per day for men – a drink being defined as one medium glass of wine or one pint of beer. Drink within these limits and we help our brains. Exceed them and in the long term we diminish our brain power – a sort of reverse supercharging. And, as I've said several times already in this book, remember: it's not what you do on one day that matters, it's the things you do every day.

The key points about stress and how to manage it are summarized in box 10.4.

BOX 10.4: MANAGING STRESS

- Managing stress well is a key element of ensuring our well-being and will improve brain power and brain health.

- Remember that there is both 'good' stress and 'bad' stress; up to a point, stress boosts brain performance.

- Consider taking up a 'contemplative activity' such as yoga, meditation or tai chi. There is considerable evidence that such activities are beneficial to well-being and have positive effects on anxiety, stress and depression as well as on brain performance.

- Exploit opportunities for humour and laughter wherever possible.

- Pet ownership, if this is possible, will reduce your stress and contribute to your well-being.

- Observing or practising any kind of visual art is proven to reduce stress levels and to affect the brain directly.

- Music – either listening to it or making it – will make you feel good and reduce stress.

- Drinking alcohol moderately, within healthy guidelines, relieves stress and is beneficial to brain health.

Darkness and light

Thanatos is the Greek god of death, his mother Nyx goddess of night, and his father Erebus god of darkness – a formidable combination. Freud said there is a Thanatos in all of us. But along with him we have also Eros, god of love; his mother Aphrodite, goddess of beauty and pleasure; and his father Ares, god of war. In our quest for well-being, we have to hold these warring forces in balance, to find equilibrium between the negative and the positive in our lives. If there is one principle I would highlight among all I have written in this chapter – and indeed in this whole book – it is that in seeking and achieving well-being we have a huge amount of control over how we order our lives to make them more enjoyable, less stressful and more productive – and in doing so, we know that not only will we feel better, we will think better too.

Covid-19 and the brain

Lions and monkeys

IN EARLY 2020, AFTER three days in a London hospital undergoing treatment for Covid-19 infection, an apparently well 55-year-old woman was discharged home. *But all was not well.* Within twenty-four hours she became confused and disorientated and began behaving oddly, going through repetitive behaviours such as putting her coat on and taking it off again and again. She hallucinated, seeing lions and monkeys prowling around the house. She became deluded, believing that an identical impostor had taken the place of her husband. She began hearing things. And she became episodically aggressive. She had become psychotic. She was just one of many patients who, in the earliest stages of the pandemic, showed apparent recovery from Covid-19 only to be struck down by serious neurological, behavioural or psychiatric disorders.

If physicians at the outset of the Covid-19 pandemic didn't expect to see neurological symptoms in what is essentially a respiratory disease, they really shouldn't have been surprised. History is littered with examples of virus-related neurology. In May 1889, one of the greatest ever pandemics started with a single case of influenza in Bukhara, in today's Uzbekistan. It spread rapidly: by November it had reached St Petersburg via the Volga trade routes, and by mid-December it was in London – that's an average of 200 miles per week. There was no air travel in 1889, but, interestingly, this evidence suggests that even if it were feasible, attempting to limit the spread of Covid-19 by restricting air travel would be ineffective. The virus will get there anyway, as long as there is some movement of people. But in the 1889 influenza outbreak, the speed of spread was not the only feature that echoes today. Variously described as nervous exhaustion, 'grippe catalepsy', 'post-grippal numbness', psychosis, 'prostration', anxiety and paranoia, the neurological effects of the virus were many. The Victorian and pioneer laryngologist Sir Morell Mackenzie described how it appeared to 'run up and down the nervous keyboard stirring up disorder and pain in different parts of the body with what almost seems malicious caprice'.[1] Neurological signs and symptoms, then, appear to be no strangers to viral pandemics. But what of brain health?

Lessons from history

Looking back into history serves a very useful purpose by helping us to put the current Covid-19 pandemic into perspective and reflect rationally on our responses to it. Our behaviour has many features in common with responses to past pandemics: social distancing, isolation, protective clothing and forced quarantines. And public disquiet – though that, it must be said, is for the most part rather innocuous compared to past public reaction. In 1771, during a plague epidemic in Moscow, the authorities provoked widespread protest by closing down the economy. Public order broke down, and the rioting mob killed Archbishop Ambrose, a proponent of quarantine.

So, before we look further at how medicine and health care have dealt specifically with 'Covid neurology', let's take a peek at how we have dealt historically with epidemics and pandemics – *for many of our modern approaches have affected our mental well-being, one of our three pillars of brain health*. If the world was not prepared for influenza in the late nineteenth century, it was similarly unprepared for the current Covid-19 pandemic emerging from China in late 2019. At present, the origin of the disease remains uncertain; all we know is that it arose in December 2019 in Wuhan, at the time home to both the infamous 'wet markets' and the military-controlled Institute of Virology[2] – and that it spread globally with devastating speed. Somewhat slower, the WHO didn't declare a pandemic until 11 March 2020.

Leaving aside the outbreak's origins and spread, how the world responded to it bears an uncanny similarity to its response to previous pandemics, notwithstanding the vast superiority of our medical knowledge over that of our forebears. In 1665, in the sleepy and ironically isolated rural village of Eyam in Derbyshire, England, a tailor's assistant, George Viccars, unpacked a delivery of wool from a merchant in London, with little apparent concern. He was not to know that the wool contained fleas – or that, unfortunately for him, the fleas were carrying bubonic plague. A few days later, on 7 September, Viccars succumbed, becoming the village's first victim – a truly deadly harbinger of what was to befall the wretched inhabitants. Once the penny dropped and the ghastly realization of what they were facing emerged, the local priest intervened, begging the people of Eyam to isolate their village in order to prevent the spread of the plague beyond its confines. By sacrificing themselves they would save others. This was not an 'easy ask'. The priest was deeply unpopular, representing the enforcement of the Book of Common Prayer on a Puritan population who had fought the monarchy with the parliamentary army of Oliver Cromwell. Now this same priest was asking them to put their lives at serious risk. But the Reverend Mompesson was very clever: he recruited their previous parson to the cause, a man who was Puritan to the core. And so they did what the two clerics asked.

Their village, retrospectively called 'the Village of the Damned',[3] was sealed off from the rest of the world. Food and supplies were dropped off at the perimeter – where warnings were erected: a veritable cordon sanitaire – and paid for by coins left steeped in vessels of vinegar. Bodies were buried where they died, in an age where being buried in consecrated ground was an article of faith. Their clothes were incinerated. Indoor church services ceased and were instead conducted in the open air, in the enigmatically named Cucklet Delf, today a beauty spot and natural amphitheatre on the edge of Eyam.

As a ghastly reward for these exceptional measures, 260 of the villagers met their demise at the hands of the illness, the last of them being farm worker Abraham Morten on 1 November 1666, bringing the fatality rate to 30 per cent.

But the parallels with Covid-19 do not end there. Consider PPE. However clever we think we are, the idea of using clothing for personal protection is as old as the hills and precedes modern knowledge of contagion. The seventeenth-century variant used widely by plague doctors in the main cities of England is shown below.

Dr Beak, a plague doctor, 1656

Invented in 1619 during the great plague in France by Dr Charles de Lorme, a physician to King Louis XIII, it consisted of a long, wax-coated leather or fabric overcoat that covered the breeches, inside which the shirt was tucked. Boots were worn – and gloves, attached to the overcoat's sleeves. Best of all, a face mask was compulsory, made of waxed leather with glass eye openings and a small beak that was filled with herbs and essential oils to purify the putrid air. We may smile at the picture; but it has to be said that the evidence for the effectiveness of our modern masks is not so solid as some may think.

Two of the very few randomized controlled trials on mask-wearing that have been conducted cast doubt on its effectiveness. The first of these, the DANMASK study, was conducted in Denmark and the results were published in March 2021. It concluded: 'The recommendation to wear surgical masks to supplement other public health measures *did not reduce the SARS-CoV-2 infection rate among wearers by more than 50%* in a community with modest infection rates, some degree of social distancing, and uncommon general mask use.'[4] The second, a study by the United States Marine Corps, goes further. It showed that strict quarantine measures, including the wearing of face masks, social distancing and handwashing, did not prevent the spread of Covid-19 in a large sample (1,848) of Marine Corps recruits.[5] However, which public health measures are effective remains a controversial issue,[6] and sifting the evidence is not helped by its heavy politicization. Indeed, there have been shocking incidents of attempted censorship of scientists.[7] For all that has been said about the protective value of wearing masks, the data are imperfect and do not warrant categoric certainty.

To me, one glaring absence has been any public explanation of *how* masks are supposed to work – an absence which I have never understood. Some simple physics is instructive here. A coronavirus particle is around 100 nanometres in size (1 nanometre is a millionth of a millimetre, and there are about 25 million nanometres in an inch – so pretty small!). The pores in the material of blue surgical masks are up to 1,000 times the size of one virus particle. In a cloth mask the gaps between fibres can be

500,000 times as big. How, then, are masks supposed to stop the movement of particles? Answer: they can't and they don't. What they can do
is to reduce the aerosol mist that passes through them.[8] But not particle
movement. As we breathe in and out in a mask, millions of virus particles
(droplet nuclei) can enter and leave. When we cough or sneeze, the mask
can stop mucus globules or spray, but it can't stop the transmission of
virus particles. When we cough, the air expelled travels at up to 50 mph
and expels an infectious mist of up to 3,000 droplets with each cough.
But sneezing beats coughing hands down, expelling air at a velocity of up
to 100 mph and carrying over 100,000 droplets per sneeze. This riveting
atomistic model has compelled a leading engineering scientist in the UK,
Dr Axon of Brunel University, to comment:

> Not everyone carrying Covid is coughing, but they are still breathing,
> those aerosols escape masks and will render the mask ineffective . . .
> The public were demanding something must be done, they got masks,
> it is just a comfort blanket. But now it is entrenched, and we are
> entrenching bad behaviour. The best thing you can say about any mask
> is that any positive effect they do have is too small to be measured.[9]

This view is supported by Norway's Institute for Public Health, which
reported that if masks did work then any difference in infection rates
would be small when infection rates are low: assuming 20 per cent asymp-
tomatics and a risk reduction of 40 per cent for wearing masks, 200,000
people would need to wear one to prevent one new infection per week.[10]
In summary, the case for wearing masks is not supported by an atomistic
model, and the advice on effectiveness is as much political as it is scientific.

The case of mask-wearing serves as just one disappointing example
of the use of evidence in the Covid-19 pandemic. In 2020, the Centre
for Evidence-Based Medicine at Oxford University lamented the absence
of high-quality, rigorous evidence to support health policy decision-
making.[11] Not a good position to be in, in the midst of a global pandemic,
either side of the Atlantic.

Lockdown: living in the shadow of Covid-19

Both in Britain and in the USA, how we have handled the Covid-19 pandemic has inflicted on many people the direst of times. Among those I despairingly include our most vulnerable groups – older citizens, young people, those in poverty and poor health (many with co-morbidities) and those with disabilities, particularly those with mental health problems. Consider the following: as we saw in chapter 6, one of the greatest influences on our brain health over the long term is our social connectedness. I've often been asked in the media about the effects of 'lockdown' on brain health. My answer has been unwavering: if you'd wanted to devise a plan to damage the nation's health (and that includes brain health) then you would have come up with social isolation. Remember the words of Charles Dickens on solitary confinement, to which many in these vulnerable groups have been subjected: 'the slow and daily tampering with the mysteries of the brain . . . immeasurably worse than any torture of the body'. Except it's not a mystery any longer. We know beyond doubt the devastating effects of prolonged social isolation on our brain health.

First, a chilling statistic: within three months of catching Covid-19 (as if that were not bad enough) about one in five people will develop a mental health problem for *the first time* – for example, depression, anxiety, insomnia, PTSD or *dementia*. These sobering data arose not from a trivial study but from a very large one – researchers at Oxford University analysed a US health database of 69 million people, of whom over 62,000 had had Covid-19.[12] The study also concluded that if you have had Covid-19, you have double the risk of developing a mood or anxiety disorder. Other studies reveal a similar picture. Britain's Office for National Statistics has reported that 'around 1 in 5 adults (21%) experienced some form of depression in early 2021 (27 January to 7 March); this is an increase since November 2020 (19%) and more than double that observed before the coronavirus (Covid-19) pandemic (10%)'.[13]

It must also be said that the risk is unfairly spread. The disadvantaged in society – vulnerable and disabled adults, and those living in economic and social deprivation – are disproportionately affected. For example, the Covid-19 mortality rate has been found to be 3.7 times higher for those under 65 living in the poorest 10 per cent of neighbourhoods compared to those living in the wealthiest neighbourhoods in England. And the explanation is unsurprising: poorer underlying health exposes us to greater risk. People in their fifties and sixties living in the poorest areas are twice as likely as those in the wealthiest neighbourhoods to have *at least two pre-existing long-term health conditions* such as heart disease or diabetes. Perversely, the principle also works in reverse: if you get Covid-19, your risk of mental illness is increased, and if you have a pre-existing mental disorder you are at twice as high risk of catching Covid. What we know about Covid-19 and mental health is summarized in box 11.1.

BOX 11.1: WHAT WE KNOW ABOUT COVID-19 AND MENTAL HEALTH

- Within three months of infection, one in five people will develop a mental health problem for the first time.

- These conditions include anxiety, depression, insomnia, PTSD and dementia.

- If you have Covid-19, your risk of developing a mental health disorder is twice that of those who are not infected.

- In the UK, the proportion of those experiencing depression has doubled since pre-pandemic times (from 10 per cent to 21 per cent).

- In the USA during the pandemic (January 2021), 38.5 per cent of adults reported anxiety disorders and 28.4 per cent reported symptoms of depression.*

- Those living in deprivation with poor health and in difficult social conditions are four times as likely to contract Covid-19 as those who are wealthy.

* Nirmita Panchal, Rabah Kamal, Cynthia Cox and Rachel Garfield, *The implications of COVID-19 for mental health and substance use* (Washington DC: Kaiser Family Foundation, 10 Feb. 2021), https://www.kff.org/coronavirus-covid-19/issue-brief/the-implications-of-covid-19-for-mental-health-and-substance-use/.

It's no wonder that the pandemic has been identified as the source of 'a growing mental health crisis'.[14] But how do we explain these risks? First, as we shall see later in the chapter, it's becoming increasingly clear that the virus directly infects the brain, and that the brain is also subject to the vast inflammatory responses that occur throughout the body. This is a story which is even now incompletely understood. But it appears highly likely that organic damage is being inflicted on the brain. Second, there are powerful psychological impacts. We know that simply being diagnosed with the disease incurs significant stress and anxiety – which can trigger depression. Covid-19 diagnosis and treatment is likely to be more traumatic than for other conditions, owing to the severity of the illness, its novelty and the associated uncertainty. And these stresses become compounded by the subsequent social isolation, in which the comfort and support of family and friends are removed. It is also known that there is an association between the impairment of the immune system and mental health.

Even though there is little to no framework for how to manage the threat to health, lifestyle and societal change posed by Covid-19, there are pointers to how we can manage the risks to our mental health during this pandemic. Some of these are shown in box 11.2. The advice there is framed around coping with the threat of infection or diagnosis. But I'd start with the wealth of advice given in the previous chapter about managing well-being. Being in a good place is a sound start, and if your well-being is in prime condition, it becomes less difficult to cope with a diagnosis. Taking care of our mental health is as important as looking after our physical health.

BOX 11.2: WHAT WE KNOW ABOUT MANAGING MENTAL HEALTH

- Limit yourself to reputable and reliable sources of information – listening to or reading 'dodgy' sources can lead to fear, anxiety and even panic.
- Keep down the hours you are exposed to news stories about Covid – overexposure to media coverage and constant monitoring of news or social media can intensify feelings of worry and distress.

- Prioritize your well-being and positive mental health – maintain a daily routine and focus on the things in life over which you have control.

- Reach out to and support friends and family – it helps us all to see the problems of others and to take a wider perspective. Talking with others is great therapy.

- Adopt the mantra 'think positive' – negativity is extremely destructive to mental health. Try to be an optimist! Focus on all those (the majority of cases) who have survived Covid-19.

- Acknowledge your feelings. Recognize that it's quite normal to feel overwhelmed, stressed, anxious or upset, especially during a lockdown. Write down your feelings in a journal; talk to others; do something creative; practise yoga or meditation.

- If you feel you are not coping or you are being overwhelmed, get help. Seek professional counselling or consult your doctor.

The pandemic reappraised

In his 1625 essay 'Of Seditions and Troubles', the English scientist, statesman and philosopher Francis Bacon presciently observed: 'The remedy is worse than the disease.' No aphorism could serve us better in questioning the coping strategies put in place by the authorities, here in the UK, in the USA and elsewhere, not least because it is arguable that they are injurious to our physical health, our mental health *and our brain health*.

To get it into perspective, we should note that Covid-19 is only one of several pandemics that have hit us in the last fifty years. Not only that, but on a global and historical scale, it's rather minor in comparison to previous devastations, with a low death rate, confined largely to older people and those with co-morbidities. Maybe you find this remark shocking; if so, consider the comparison of death rates across a series of historical pandemics by the statisticians of the *Financial Times*, set out in table 11.1.

TABLE 11.1: GLOBAL DEATH RATES ('CRUDE' MORTALITY RATE: % OF AFFECTED POPULATION)

Disease	% deaths
New World smallpox	93
Plague of Justinian	55
Medieval Black Death	39
Spanish flu	2.4
HIV/AIDS	0.7
Covid-19	0.03

Note: The 'case fatality rate' (CFR) would be much higher in these examples, as the CFR is related to the number of diagnosed cases – and of course not everyone catches the disease. The actual risk of dying if you or I catch Covid-19 is the infection fatality rate. This statistic is probably what we all want to know. Unfortunately, it is notoriously difficult to calculate, because it depends on the total number of those who get infected – and of course we don't know that either. Many people are asymptomatic and we don't have universal testing.

Source: Anna Gross, 'From plague to polio: how do pandemics end?', *Financial Times*, 11 March 2021, https://www.ft.com/content/4eabdc7a-f8e1-48d5-9592-05441493f652.

uncommon. On 23 April 2021, a leading US newspaper proclaimed: 'The total number of Covid-19 deaths so far is on track to surpass the toll of the 1918 pandemic which killed an estimated 675,000 nationwide.'[15] What the journalist did not point out was that in 1918 the US population was only 103 million, whereas today it is over three times that figure at 333 million. Adjusted for population size, the 1918 pandemic would have killed over 2 million today – nearly four times as many as US deaths from Covid-19. As at 26 July 2021, there had been 610,891 Covid-related deaths in the USA (1 death per 545 people), and in the UK there have been 129,418 deaths in a population of 68 million (1 in 527 people).

Nor is it only relatively low death rates that should help us to rationalize our view of Covid-19. At the time of writing, the latest projection is that only 1 in 200 people worldwide will have suffered an infection.[16] In the USA, there have been 34.44 million cases (i.e. one in ten people have caught Covid-19) and in the UK, there have been 5.69 million (one in twelve people).

It's therefore entirely reasonable to ask whether the extent of the measures that have been taken to curb the pandemic – which together have made serious inroads into public liberties, the economy and health – are justified in relation to the severity of the disease. We already know that there has been a substantial impact on mental health, but we don't yet know the longer-term health consequences, and those will include an impact on brain health. Let's now look at that impact.

The natural history of Covid-19

Covid-19 is a respiratory disease caused by infection with the virus SARS-CoV-2 (Severe Acute Respiratory Syndrome CoronaVirus 2).

SARS-CoV-2 virus particles in droplets

Covid-19 symptoms include high temperature, a persistent dry cough, headache, debilitation, difficulty with breathing, and loss of smell and taste. Up to fourteen days – but more usually four to five days – may elapse between infection and the onset of symptoms, and this is a very dangerous period. The majority of Covid-19 cases are caused by infected, pre-symptomatic people. Only about one in five people remain asymptomatic, and they are far less infectious than those with symptoms.[17] Symptoms in those who do become unwell are mild to moderate for most (over 50 per cent), severe for about 15 per cent – including, for example, dyspnea (difficulty in breathing) and hypoxia (low oxygen availability) – and critical for about 5 per cent (respiratory failure, shock or multi-organ dysfunction). After the onset of symptoms, people are infectious for up to ten days in moderate cases and up to twenty days in severe cases. The biggest risk factors for fatality from Covid-19 are age, maleness (men face twice the risk of women) and co-morbidity. And in some, a range of effects ('long Covid') persists for months after recovery from the acute stage.

Contagion occurs when we are exposed to virus-containing *respiratory droplets and airborne particles* exhaled by someone who is infected.[18] The virus may be inhaled, entering via the mouth or nose, or may enter via the surface of our eyes, either directly or indirectly through touching. The risk factors for infection include how long we are exposed to the source, how far we are away from it, how crowded the environment is and the exchange rate of air (ventilation). In overcrowded, unventilated conditions, virus particles can remain suspended in the air for hours. Touching a contaminated surface is now thought to constitute a low risk of transmission.[19]

The body's response to Covid-19 infection

You may be asking, why is getting Covid-19 such a terrible and even life-threatening experience for some people? A report in *Nature* sums up its seriousness somewhat alarmingly: 'Disease severity in patients is due not only to the viral infection but also the host response.'[20] Let me paraphrase this. The deadliness of the infection *is attributable as much to the aggressiveness of our own immune system as to the effects of the disease itself*. We may unleash upon ourselves an inflammatory response of such severity that it threatens our recovery and even our life. What is this response? And what does it do to our brain health?

When we breathe the SARS-CoV-2 virus into our lungs, a protein on the surface of the virus (the 'S-spike') binds on to tiny patches of protein on the surface of our lung cells (the alveolar epithelial cells). These tiny patches of protein are called ACE2 (angiotensin converting enzyme 2) receptors. They are the virus's gateway into the cell, like a key in a lock. (For the curious among you, ACE2 receptors are not there for the convenience of the virus – they are part of a vital system for the control of blood pressure, wound healing and inflammation.) ACE2 receptors are abundant in the lungs but are not confined to them. They have been found in other organs such as the heart, the liver and kidneys, blood vessels, the eyes and nose, *and the brain*.[21] Interestingly, ACE2 also floats around in the blood as a soluble form. Here it is able to bind with virus particles and neutralize them. Children have high floating reservoirs of ACE2 in their plasma, and this may be one reason why Covid-19 infections in children are infrequent and minor. Conversely, soluble plasma ACE2 is present only at much lower levels in older people, who are at greater risk of severe illness.

Once the virus enters the cell, our immune system swings into action. If you are healthy, SARS-CoV-2 infection and the destruction of lung cells triggers a local immune response. White blood cells including macrophages and monocytes are recruited, substances called cytokines are released, and adaptive T and B immune cells are primed. In most cases, this process is capable of resolving the infection. But if you are unhealthy (if, for example, you are suffering from long-term illnesses such as obesity or diabetes) and/or you are older (say, over 75) and/or your immune system is impaired, or if your infection results from a huge viral load, then a very different story may unfold. There is likely to be an *uncontrolled immune response* – the release of a large dose of inflammatory molecules (a 'cytokine storm'), a massive influx of immune cells to the infected area and a failure of the body to inhibit pro-inflammatory responses.[22] Up to twenty different inflammatory cytokines have been found in the blood during severe cytokine storms. There is a buildup of immune cells and cellular detritus (rubbish) in the lungs; lung tissues experience swelling, damage and fluid buildup (pulmonary oedema). In the early stages of infection the greatest clinical concern is hypoxemia (low blood oxygen), which gives way to ARDS (acute respiratory distress syndrome) and eventually respiratory failure (and sadly, for many, death). But it's not only the lungs that are affected: many other organs are damaged, both directly because of low oxygen, and indirectly by their own local inflammation and blood clotting – including the brain.

Covid-19 and the brain

SARS-CoV-2 is said to be a respiratory virus. But its earliest symptoms are not respiratory. *They are neurological.* We are talking about headache, loss of smell and loss of taste. Moreover, delirium – confusion, with reduced cognition and memory – is often the only presenting symptom, even in younger patients. Its incidence in severely ill Covid-19 patients in intensive care units is as high as 84 per cent.[23] About 80 per cent of Covid cases, many of them minor, reveal neurological or psychiatric symptoms, usually after recovery from Covid-19 itself.

Patients have been seen with encephalitis (brain inflammation), with seizures, with imbalance and with paraesthesia (tingling sensations), and some people with Covid-19 briefly lose consciousness. There are blood clots, strokes and TIAs (transient ischemic attacks or 'mini-strokes'). There is anxiety, depression and even psychosis. How can these effects on the brain be explained?

In September 2020, a news article in *Nature* stated: 'Although viruses can invade and infect the brain, it is not clear whether SARS-CoV-2 does so to a significant extent.'[24] A few months later, in January 2021, *The Scientist* ran a leader article which announced: 'Autopsy studies have yet to find clear evidence of destructive viral invasion into patients' brains.'[25] This state of uncertainty posed some difficult questions, not least among them: how do we explain the increasing prevalence of neurological and psychiatric conditions? But a mere five months *and an estimated 20,000 scientific papers* later, there was among that body of work sufficient evidence not just that the Covid-19 virus does enter the brain, *but how.*

The intriguing picture that science has revealed, in this startlingly short time-span, looks like this:

- The SARS-CoV-2 virus *can enter and infect the cells of the brain*, including neurons, but predominantly it enters immune cells and cells of the lining of the blood vessels (endothelium). The virus enters through the highly impermeable blood–brain barrier (BBB) – the tightly sealed blood vessels which serve and protect the brain. The virus spike locks on to the ACE2 receptors of the endothelium, replicates inside its cells and then passes into the brain. Or it may enter through an ACE2 docking site and pass directly through the endothelium as a 'vesicle' (a little fluid-filled bladder) without replicating itself.[26] There is some good news, however: it appears that though the virus gets into neurons, it doesn't kill them;[27] nor are white matter cells demyelinated.[28]

- Other evidence has shown that a key 'brick in the wall' of the BBB, the defence cells we call astrocytes, also have ACE2 sites and are penetrated by the virus, allowing it to enter the brain, acting as a sort of 'Trojan horse'.[29] As the BBB becomes progressively infected, its permeability increases, a sort of vicious circle in which the BBB cells become weaker and fail.

- When infected, the astrocytes set up local inflammation, releasing cytokines and chemokines which can enter the brain.[30] The BBB will also react to the massive general levels of inflammation set up by the virus in the body. Soon, the brain becomes deprived of essential nutrients as the transport systems in the BBB break down. Its communication with the immune system also stalls, and this is especially damaging. It's interesting and not widely known that inflammatory damage like this to the BBB generally causes many complaints unrelated to Covid-19, including fever, discomfort and malaise, memory impairment and post-sickness depression.[31]

- SARS-CoV-2 introduced into the noses of mice results in its appearance in their brains,[32] and in humans the virus has been found both in the lining of the nose and in the brain stem.[33] These findings clearly establish that the virus can enter the brain, but do not tell us whether the virus is taken up by the nasal nerve cells, is absorbed via blood vessels in the nose, or comes from an infected lung which then seeds the brain via the general blood circulation. Paradoxically, it seems that the virus may well not enter nerve endings in the nose even though it may deprive us of our sense of smell.

- In yet another Covid paradox, startlingly low oxygen levels are found in patients who are not short of breath. The resulting hypoxemia not only deprives the brain of necessary oxygen but further ramps up the 'cytokine storm'. And, in a surprising twist, infection of the respiratory centre in the brain stem may well be a cause of ARDS and respiratory failure.

- About one in five Covid-19 patients show cerebral vascular events, such as strokes, TIAs and blocked blood vessels, as a result of circulating micro-embolisms (travelling blood clots) and increased clotting factors in the blood.[34]

- The lungs are ground zero; but it's been found that the virus will also attack just about everything else with an ACE2 receptor in it – liver and intestines (10–15 per cent of cases), heart (20 per cent), kidneys (75 per cent) and testicles (less than 5 per cent), causing the multi-organ deficits characteristic of Covid-19. Interestingly, this doesn't happen much in the immune system, where ACE2 receptors are rarely found. This multi-organ damage is rarely seen with other viruses.[35]

To sum up, then, the Covid-19 virus appears to contribute to neurological problems in two ways – it enters the brain and infects some of its cells; and it generates severe and damaging local and systemic inflammation. A summary of what we know (or think we know!) about Covid and the brain is shown in box 11.3.

BOX 11.3: WHAT WE KNOW ABOUT COVID AND THE BRAIN

- SARS CoV-2 is a respiratory virus but the first symptoms are neurological, including loss of taste and smell, headache and delirium.

- The virus can enter and infect cells in the brain, mainly immune cells and blood vessels.

- The blood–brain barrier is breached by the virus and damaged in the process, producing symptoms of discomfort and malaise.

- The virus can enter the neurons but does not appear to kill them, nor does it damage the myelin sheaths of communicating fibres.

- The brain is subject to local and systemic inflammation with serious and lasting consequences.

- The effects of Covid-19 infection may be prolonged and last for months or years.

- There are over 200 recorded symptoms of 'long Covid', many of them neurological, behavioural and psychiatric.

- How 'long Covid' should be treated is under investigation and development.

Long Covid

Young, never hospitalized and, some would say, neglected. The 'long-haulers'. Between 10 and 20 per cent of all cases develop into 'long Covid', where people become seriously debilitated for over six weeks after their infection has ended. While two-thirds of those with 'long Covid' say their ability to carry out their activities of daily living is impaired for twelve weeks or more, many will go on to have symptoms for over twelve months. Findings from the UK's Covid Symptom Study show that around 10 per cent have had symptoms for thirty days and 1.5–2 per cent for ninety days, and that 'long Covid' appears to be twice as common in women as in men.[36] Drawing on what we already know about post-viral fatigue, if you're not better within a year, you may well not get better for years or decades. And the numbers are substantial – in July 2021 in the UK it was projected that 3,000 people per day were finding themselves incapable of working for three months or longer.[37] The commonest symptoms indicating 'long Covid' are fatigue, shortness of breath, chest pain or tightness, 'brain fog' (problems with memory and concentration), difficulty sleeping (insomnia) and heart palpitations.

Even a cursory look at 'long Covid' reveals the largely neurological and psychiatric nature of the condition. A study published in May 2021,[38] looking at patients who had suffered Covid-19 symptoms for longer than six weeks, showed that many had neurological symptoms: fatigue (85 per cent), 'brain fog' (81 per cent), headache (68 per cent), numbness/tingling (60 per cent), dysgeusia – loss of taste (59 per cent), anosmia – loss of smell (55 per cent) and myalgias – muscle pain (55 per cent). Interestingly, none of the 100 patients in this study had required hospitalization for respiratory illness during their acute period. The authors of the study emphasize the extent to which patients had had their quality of life ruined by prominent and persistent brain fog and fatigue.

Brain fog – and brain ageing

Feeling confused, fatigued? Thinking slowly, having trouble in concentrating and remembering? All are common symptoms of 'brain fog', a non-medical term used by people when they feel 'spaced out', sluggish or fuzzy. It's commonly experienced as a result of stress, sleeplessness, hormonal changes, medical conditions such as multiple sclerosis and mild cognitive impairment – and now it is increasingly reported after infection with Covid-19. Are we able to explain brain fog and how can it be resolved? As we've seen, there's now substantial evidence that in a Covid-19 infection the brain is swamped by cytokines, becomes depleted of oxygen and nutrients, and is directly invaded by the virus. Brain fog as an outcome should not, then, surprise us. But there is more. One possible explanation is that the virus damages the mitochondria of our brain cells.[39] The mitochondria are the powerhouses of our cellular energy and they require high oxygen levels – so they are ideal targets for a replicating virus. Impairing the mitochondria and thus our cerebral energy levels will undoubtedly subdue our cognitive processes and result in 'brain fog'. But within this deadly mechanism there is a cunning plan. It's a powerful evolutionary weapon in the armoury of the SARS-CoV-2 virus. The resulting cognitive impairment helps the virus to spread, as infected individuals are far less capable of anti-infection behaviour.[40] Fortunately, on a more optimistic note, recent research shows that the mitochondria are extremely resilient. They have a double system of repair: via their own internal mechanisms and via help from the cell nucleus. Studies have shown that nutrition – and some would say nutritional supplements – are key to the support of these natural repair processes.[41]

If we needed any further encouragement to diminish the risks incurred through infection with Covid-19, then consider this: a joint study by British and US researchers has linked severe Covid-19 cases to mental decline equal to ageing a decade.[42] Though this study has yet to be peer reviewed, it's in line with the widely accepted finding that cognitive decline is common after infection.

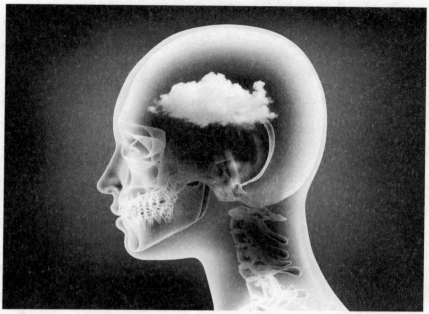

Brain fog

Practical recommendations

In summary, Covid-19 is a highly infectious disease inflicting serious harm, particularly in certain ethnic groups, older population groups and people with co-morbidities, in numbers such that the integrity and working capacity of our health services are threatened. By contrast, Covid-19 is generally much less virulent in younger, fit individuals, especially children. And to remind ourselves of another Covid paradox: though in its contagion it can be ferocious, across the population its lethality is low. Many of its infectious features are unique. One of these is its capacity, as a respiratory virus, to inflict damage on a wide variety of organs – including the brain, where, as we have seen, it is responsible for a wide range of persistent neurological behavioural and psychiatric problems. So, as well as everything else, Covid-19 is a real threat to brain health. What, if anything, can we do about it?

Advice on reducing the risks of a Covid-19 impact on brain health falls into three categories:

1   Avoid infection.

2   Protect your immune system.

3   Take specific measures to look after your brain health.

Avoiding infection

The main routes of infection are shown in figure 11.1.

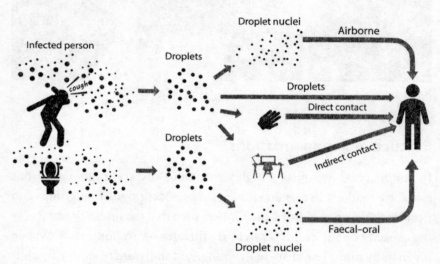

Figure 11.1: Main routes of infection for respiratory diseases such as Covid-19

Note: 'Droplet nuclei' are viral particles.

The best non-pharmaceutical interventions are ventilation, social disancing and – with some caveats – masks. But we should also note one route little mentioned: infection via faecal matter. We'll come on to that in a minute.

First, ventilation. The vast majority of infections occur indoors, and there are more infections in poorly ventilated spaces.[43] The good news is that coronavirus particles are very easily removed by air extraction because of their small size. The bad news is that most air conditioning simply recirculates indoor air and exposes everyone to airborne pathogens, unless a high-efficiency particulate air (HEPA) filter has been used. Fresh air appears to be king. For practical purposes, the air flow through any indoor space (including aircraft cabins, public transport and cars) should be increased by the use of extractor or blower fans, by opening windows and doors where possible, or by using filtered air conditioning. Closed, unventilated, indoor spaces are deadly. If increasing ventilation isn't possible, then the risk should be mitigated by other means, for example by an effective mask (see overleaf) or by reducing the number of people sharing the space. In passing, we should mention the Amoy Gardens incident in 2003. In this high-rise building in Hong Kong, the SARS virus spread because the U-bend systems in the lavatories dried out and allowed the airborne dispersion of virus particles. Not only were the particles breathed in; they also settled on surfaces throughout the apartments, and contagion occurred through touch and indirect means. The advice is clear – do not ignore foul smells in bathrooms, kitchens or washing areas. Investigate and fix! But it's more basic than this. If a person is infected, especially if they are older, its likely that the virus will be present in bodily excretions. As we flush the toilet, an aerosol plume immediately leaves the bowl and enters the air, where of course it can be breathed in – or, as in the Amoy Gardens, settle on the surrounding surfaces. The advice is simple: close the lid before flushing, keep surfaces clean, wash hands and ventilate the lavatory!

Second, social distancing. It's now believed that the Covid virus can spread either in droplets smaller than 5μm,[44] or as particles which we know may remain airborne in a droplet cloud for up to 8 metres (nearly 9 yards) from the source.[45] Talking spreads more droplets than merely breathing, and the more loudly we talk, the more droplets are expelled. For example, one minute of loud talking expels more than 1,000 infectious droplets. An interesting hypothesis therefore arises: those who talk (or sing) the most and the loudest may be super-spreaders! It's known that

the risk of inhaling virus particles or airborne droplets decreases with distance from a source, because droplets will fall to the ground as they travel
and particles disperse.[46] Indeed, if we are standing further than 2 metres
from an infected person, the concentration of virus particles emanating
from them drops off dramatically. Though distance is only one factor
(others being temperature, light and humidity, all of which affect the via-
bility of the virus), the advice to 'keep your distance' is well founded.

Third, masks – a controversial issue which we have already discussed.
To get the best out of a mask, it should be made of a material which filters
virus particles between 60 and 140 nm, and it must be worn correctly –
which is not a simple matter: the Mayo Clinic lists a twelve-point protocol
for proper mask-wearing. What we do when wearing a mask, too, is
doubly important – it changes the effectiveness of the mask, and it may
increase the risk of transmission because of the tendency for wearers
to touch their faces *and* vent the mask when adjusting it.[47] Surgical and
N95 medical-grade masks are the most effective in protecting against
infection,[48] but even then the N95 mask requires sealing to the face with
Vaseline to optimize its performance.[49] In summary, wearing a mask
reduces the risk of infection but does not eliminate it. Wearing a respir-
ator with a HEPA filter gives more complete protection.

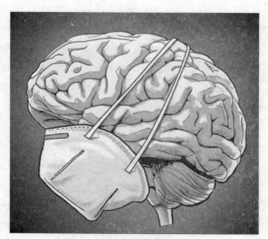

Masking the brain

Before we leave this section, let's look at *la pièce de resistance*: vaccination. Vaccination is highly effective and provides very high levels of protection – over 90 per cent in most cases. But it isn't immediate: there is a latency period of five days after receiving the shot (first or second) in which we can still become infected. Even so, the severity of the disease, the chances of hospitalization and even death are greatly reduced for the vaccinated. It's still possible to spread the virus after vaccination, but this possibility is very low. Vaccination is inherently a self-protection measure. We are not sure, though, how long the vaccines' protective effects last. Follow-up annual vaccination is the likely future policy as the disease stabilizes. And this brings us to a vital question: how and when do pandemics end?

Amazingly, this question was answered over 350 years ago by John Graunt, widely recognized as the father of epidemiology, as he pored over parish mortality records after years of the plague, typhus and smallpox. His conclusion? A relatively simple one – the disease doesn't go away; deaths simply revert to normal levels, what we call today the reduction of 'excess deaths'. One measure of this normalization is that there is no longer an unsustainable threat to the provision of health care. Covid-19 isn't going anywhere soon. But it will become *endemic* (a disease regularly found in a certain region or area). At the time of writing, it appears that we are approaching endemic levels – cases might be up but hospitalizations are down, as are deaths. Let's look at the evidence for the UK. At the peak of the first wave (14 April 2020) there were 939 deaths per day. At the peak of the second wave (23 January 2021) there were 1,235 deaths per day. On 24 June 2021 (in the third wave) – by which time over 70 million doses of vaccine had been given – *there were 'only' 65 deaths per day*. Now let's look at hospitalizations. In Britain in January 2020, a daily case number of 100,000 would have meant 10,000 hospitalizations – a ratio of one in ten. In July 2021, after some 60 per cent of the adult population was fully vaccinated (having had two shots), the ratio was between one in forty and one in fifty. And these hospitalizations were largely of unvaccinated young people, who require much less intensive treatment and are discharged earlier. Vaccination has truly been a game-changer. It has broken the link between infection, hospitalization and death. This **apparent defeat** of an infectious disease by prophylaxis, vaccination and anti-virals is a triumph of medical science which we should celebrate.

Finally, let's examine a common fear – that our current vaccines will not be effective because of genetic changes, called mutations, arising in the virus. This fear is exaggerated. First, while viruses do indeed constantly change, most of the time these mutations are of no consequence, because one adjustment doesn't change the protein configuration of the virus spike. These small changes are called 'drift' (as opposed to major changes, which we call 'shift'). Second, all current vaccines are developed (as future ones will be) to cope with genetic drift, and have been tested on many variants to ensure their effectiveness. The antibodies created after vaccination will recognize and respond to many variations of the virus spike protein. And we know that infectious agents such as viruses usually weaken over time. Currently, researchers and vaccine developers are working on robust vaccines that will respond to new threats.

Protecting your immune system

Good general health is not only good for the brain, it also offers protection against severity of infection – and for the same reason: if you are in poor health (if, for example, you have pre-existing chronic health conditions), then your immune system will be weakened and your levels of inflammation are likely to be higher. And, as we saw in chapter 1, inflammation is the enemy of good brain health.

If we become infected by the SARS-CoV-2 virus and we already have high levels of inflammation, our immune response is likely to go into overdrive, resulting in a very damaging cytokine storm, the results of which are as harmful as the viral infection itself. One of the principal reasons why age is such a risk factor for Covid-19 is that the immune system becomes impaired as we get older. One sign of that is that our immune system responds less well to vaccination as we age. Looking

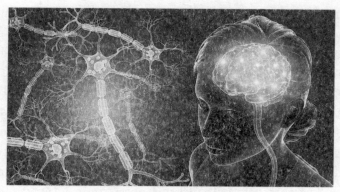

Inflammation

after our immune system is therefore critically important. How can we achieve that?

Let's start with nutrition. Simply put, and contrary to the many claims made in advertising and the media, we cannot 'boost' our immune system through diet, and no specific food or supplement will prevent us from catching Covid-19. However, it's certainly true that there are many nutrients necessary for the normal functioning of the immune system, and we should make a special effort to consume them. They include vitamins A, B6, B9 (folate or folic acid), B12, C and D. Important minerals are copper, iron, selenium, magnesium and zinc. You can look up most of these and their availability in chapter 5. Modern diets are often deficient in nutrients because, on both sides of the Atlantic, we eat a woefully homogeneous diet. Over 75 per cent of all modern food is derived from only five animal and twelve plant species. Astonishing! Diversifying the range of foods we eat will greatly reduce the need for supplementation – and most nutrients such as vitamins and minerals are required in very small amounts. If you still need convincing, remind yourself of the advice of both the British NHS and the US NIH: there is insufficient evidence for the use of any dietary supplement – whether vitamin, mineral, herb or other botanical, fatty acid or any other – as a means to prevent, lessen or treat Covid-19. But strengthening our immune system by improving our diet is quite another matter.

We should remind ourselves at this point that over 70 per cent of the immune system is found in our gut – and for a very good reason: vast amounts of potentially toxic and pathogenic material may enter the body through the food we eat – including the SARS-CoV-2 virus. It's vital to keep your gut microflora diverse, healthy and well balanced. These micro-organisms stimulate the production of immune cells (such as lymphocytes – particularly T-cells) and antibodies. It pays to look after them! And not only for this reason: as we saw in chapter 4, the bugs in our gut have a powerful system for communicating with the brain, and provide it with psychotropic molecules such as serotonin. Chapter 4 includes a full range of tips on looking after gut health.

Beyond doubt, one of the best ways to fortify our immune system is through physical exercise. We know that physically active, lean individuals have reduced resting levels of inflammation. Regular exercise has an overall anti-inflammatory influence. Moreover, regular exercise counters many elements of the disease process in humans – for example, it decreases oxidative stress, promotes the circulation of key immune cells and, most reassuringly in the midst of a pandemic, improves the surveillance of our immune system. Hurrah! But still – a double caution is in order. First, we shouldn't overdo it. It's known that high-intensity training for more than sixty minutes temporarily depresses our immunity and puts us at increased risk of infection – reflecting the sheer amount of overall stress we are imposing on the body. Second, as we saw in chapter 2, one session of purposeful exercise every day is not an antidote for daily sloth, just as it is not an antidote for overeating or smoking. The message boils down to this: be more active every day, sit less, and exercise regularly and moderately. Above all, balance your food intake and your activity levels to avoid any weight gain.

By social distancing, we place ourselves in an appalling dilemma. It undoubtedly lowers our risk of contracting Covid-19. *But it also weakens our immune system.* Why should we doubt this? One of the biggest factors in beefing up our immunity is to have contact with foreign pathogens, a beautifully poetic paradox if ever there was one: to become resistant to disease, get infected. As the old saying goes, 'What doesn't kill you, makes you stronger.' As any paediatrician will tell you, you should allow your child to get cuts and scrapes, mix with other infected children, and get coughs and colds. One of the best killing machines the Europeans brought to the New World was infectious disease, notably smallpox: it killed some 50 million unprotected native people. But there's a more subtle point to be made here than just good old biological immunity. Social isolation wears down our health, our resilience and our ability to cope. New evidence shows that via numerous mechanisms (such as the hypothalamic–pituitary axis, discussed in chapter 6), social isolation will suppress our immune responses, *just when we need them the most*. Above all, social isolation is highly stressful, and stress persistently and seriously disables the immune system.[50]

To all those who have asked my advice on how to lessen the impact of social distancing and social isolation, I have given the same answer: use every means possible to retain human interaction with friends and family. Write letters, send photographs, make phone calls, use email and other social media – anything to maximize your contact with others.

Finally, something a lot of us will want to know: does sexual activity strengthen the immune system? As we saw in chapter 7, not only will lifelong, regular, rewarding sexual activity improve feelings of well-being, but habitual sexual intimacy, especially with an emotionally close partner, confers benefits in high-level thinking skills. These include memory and recall, mathematical performance, spatial awareness and verbal fluency. Wow! Now new research findings have shown that our 'innate' immune system is activated by sexual arousal and orgasm.[51] In an experiment at the Essen University Clinic in Germany, young volunteers masturbated to orgasm and their blood samples were compared to those of participants in a non-sexual activity. Sexual arousal and orgasm increased the absolute number of leukocytes, in particular natural killer cells. By contrast, T cells and B cells, as well as interleukin 6, were unaffected by sexual activity. And there are indirect effects via the copious release of feel-good hormones, such as serotonin and dopamine, which suppress the production of cortisol, an antagonist of natural immune responses. The effects of stress or its absence may well explain a fascinating experiment undertaken at Wilkes University in America, where scientists investigated the effect that sex had on levels of IgA (immunoglobin A). IgA is an antibody and an important first line of defence in our resistance to infection. It inhibits bacteria and viruses from sticking to epithelial cells, and it also neutralizes bacterial and viral toxins, both inside and outside our cells.

The results of examining the salivary IgA levels in over 100 under-graduates showed that those who had sex less than once a week had

Sex and Covid: the lovers' dilemma

very slightly higher levels of IgA than those who abstained completely – but that those who had one or two sexual encounters each week had levels 30 per cent higher. And what of those who had very frequent sex? Surprisingly, those who had sex more than three times a week had *lower* IgA levels than the abstainers. These results are open to interpretation. But it looks as though the answer may well lie in relationships – stress is lower in those with a settled relationship, but in those students who were highly sexually active, the researchers surmised higher levels of anxiety with many partners. And we know that stress and anxiety depress IgA levels.

There can be only one conclusion: it's reasonable to say that moderate levels of sexual activity in a close relationship are beneficial to both the brain and the immune system.

Looking after your brain

So now the $64,000 question: knowing what we do about the SARS-CoV-2 virus, are there any specific measures that will benefit our brain health? One important area is the integrity of the blood–brain barrier (BBB). The first thing to say is that the BBB is extremely sensitive to ageing: the way it works changes as we live out our years. Having said that, it's also true that the BBB can be older *and* healthy – it's just different with age, so we shouldn't fear that just because we are older, the BBB will be disrupted. Where we see cognitive decline, however, there's evidence that it is accompanied by BBB leakage.[52] Most age-related changes are non-genetic, but not all of them. About 25 per cent of us have the ApoE4 allele (gene), which is a risk factor for Alzheimer's disease. The ApoE4 allele can affect BBB integrity during normal cognitive ageing. Recent work suggests that the leak in the BBB that occurs with Alzheimer's may be attributable to the loss of certain BBB cells called pericytes. Astrocytes, by contrast, seem to be overactive in people with Alzheimer's. This is where it starts to get interesting. It might be possible to preserve the health of the pericytes by using some of the same interventions that extend lifespan, such as regular exercise, calorie reduction and improved diet. Specific nutrients which support the BBB or are known to protect neurons include astaxanthin (found in salmon, red trout, lobster, crab and prawns), curcumin (a polyphenol found in turmeric), sulforaphane (found in cruciferous vegetables such as broccoli, cabbage and Brussels sprouts), resveratrol (found in grapes, red wine, dark chocolate and pistachio nuts), omega-3 fatty acids, and Vitamins B1, B12 and D. You will often find these nutrients sold as supplements, because it is argued that it's difficult to absorb enough of them from natural sources. Some, such as astaxanthin, are marketed as 'anti-ageing' treatments, but they are most definitely not 'snake oil'. All of them have well-documented effects on our metabolism and are non-toxic. For example, resveratrol improves cognitive function and moreover has been used therapeutically as a successful anti-viral treatment – indeed, it has been suggested as a treatment for Covid-19. If taken orally it has 'low bioavailability' (i.e. is not easily absorbed into the blood, so its plasma levels after consumption are low). This problem is overcome if it is introduced nasally. Resveratrol inhibits SARS-CoV-2 replication by 98 per cent, slows down viral entry to the cells and, if given early, can prevent uncontrolled inflammation. A clinical trial has been proposed.[53]

And then there are mitochondria. We've already mentioned the evidence that Covid-19 interferes with the mitochondria in our neurons. Is there anything we can do to preserve mitochondrial function? The answer is 'yes', and all the following measures are backed up by research:

- *Lowering calorific intake* (but not nutrients) reduces the output of reactive oxygen species which drive inflammation and cellular damage, including in the mitochondria. Aim to establish a healthy calorie range to prevent overeating – a reduction of about 10 per cent is recommended (see chapter 3).

- *Intermittent fasting* (on which also see chapter 3) supports the mitochondrial network by removing damaged mitochondria and triggering biogenesis (manufacture) of new mitochondria.

- Astonishingly, certain neurons in the brain 'feel' a sudden rise in glucose levels (a 'sugar rush'). Their mitochondria rapidly change their shape and structure, and this can lead to profound overall metabolic change such as type 2 diabetes.[54] *Avoiding regular sugar 'boosts'* from the consumption of refined carbohydrates and sugars will remove this pressure on our precious mitochondria.

- By contrast, *you can consume foods containing nutrients which support healthy mitochondria*, for example by boosting their enzymes. Examples are resveratrol, omega-3 fatty acids (cold-water, fatty fish), alpha-lipoic acid (grass-fed red meat, liver, spinach, broccoli), and L-cartinine and creatinine (lean red meat, poultry, eggs, nuts, beans, seeds). See chapter 5 for more details.

- Force the mitochondria to generate energy by *exercising regularly* (see chapter 2). This is especially important as we get older, because there is an incipient loss of mitochondria as we age (1–2 per cent per year from middle age onwards). Exercise will reduce this trend.

- *Try heat therapy*, such as regular saunas – two or three per week for 10–15 minutes at a time. Research has shown that increasing the temperature of muscle tissue increases the efficiency of mitochondria.[55]

- *Getting a good night's sleep* is about more than not feeling tired. Hormones which act to regulate our mitochondria, like corticosteroids, are extremely sensitive to the disruption of our daily rhythms.[56] Look after your sleep, look after your mitochondria!

- *Relaxation* and *meditation* can play a part in keeping our mitochondria healthy by reducing the psychological effects of stress hormones, such as cortisol.

Finally, one really fascinating line of research – essential oils. Two of the major types of active ingredients in essential oils are terpenoids and terpenes, whose molecules are exceptionally small and arrive in the nasal cavities in extremely high densities. They can transfer across the nasal lining if inhaled, or penetrate the skin after topical application, and thereafter enter the blood and cross the BBB.[57] A survey of available research shows that aromatic plants and essential oils are active against a large number of viruses, including herpes 1 and 2, HIV, adenovirus, hepatitis B, enterovirus, the Junin virus that causes Argentinian haemorrhagic fever – and even against SARS-CoV-1, which has 96 per cent of the same genetic background as SARS-CoV-2.[58] Attention has been drawn to one particular essential oil – Carvacrol, a monoterpenoid found commonly in many botanicals, such as thyme, oregano, black cumin and wild bergamot. Among other monoterpenoids, it has been reported to inhibit ACE2 activity, block the entry of the SARS-Cov-2 virus and bind specifically with M[pro], an important enzyme in the viral genome. Many well-conducted, rigorous studies have highlighted its anti-viral, anti-inflammatory, anti-oxidant and immune-moderating properties.[59]

A summary of these recommendations is given in box 11.4.

BOX 11.4: COVID-19 AND BRAIN HEALTH – RECOMMENDATIONS

1. Avoid infection
- Take all necessary precautions to reduce the risk of infection.

- Consider getting the vaccine as soon as you are able.
- Understand what to do if you think you may be infected.

2. Protect your immune system
- Maintain a healthy diet.
- Look after your gut health, including good oral hygiene.
- Exercise regularly *and* increase your daily activity levels.
- Stay socially connected, including good intimate relationships.

3. Take specific measures to look after your brain health
- Follow the special measures applicable to the brain beyond those of general health.
- Take care of your sense of well-being and pay specific attention to your mental health.
- Strengthen the blood–brain barrier.
- Look after your mitochondria.
- Monitor your brain health, including paying attention to any sudden confusion.
- Research is advancing rapidly – read widely to identify new helpful research.

Conclusion

As I conclude this chapter, I am only too aware of the rapid scientific advances that will add an element of redundancy to it, as new science overtakes old, even if some of it was published only a few weeks or months ago (I write in July 2021). I have already mentioned the astonishing breakthrough earlier this year, when in May 2021 scientists answered a question that was unanswerable in January (Does Covid infect the brain?). We should rejoice in such progress. Rarely does our scientific knowledge progress at such a rate and in such a way as to rescue humanity so promptly. It goes without saying that we have much yet to learn.

But I am reminded acutely of an incident in 2006 as I sat observing the council meeting of an international medical conference in Salzburg. This meeting was addressed by its president, Bruno Lunenfeld, an eminent Israeli professor. He stood before his august body of experts and, after pausing to look searchingly around the table, made a sanguine and prophetic statement: 'My dear colleagues,' he began, 'our only hope for the future of human health is science.' How true then. And how true today. Though the long-term effects of Covid-19 will be with us for some time, our hope must lie in the awesome power of science to defeat and override its pernicious effects on the health of humanity.

Epilogue

CHARLES EUGSTER IS ONE of my greatest heroes. You may have never heard of him, but he is a salutary example of what this book is all about, maintaining his brain health and thinking power into his late nineties. At the age of 85 *he began a fitness programme*. It was, in his own words, 'purely vanity – I looked in the mirror and didn't like what I saw'. Eleven years later, he won two gold medals at the World Masters Athletics Championships in Lyon – and in 2012 he gave an astonishing TEDx talk in Zurich on 'Why bodybuilding at age 93 is a great idea'. It was a compelling, electric performance, drawing in the audience and engaging them in gales of laughter, admiration and amusement. An incontrovertible display of true brain power.

What lesson is there for us in this tale? When I was in my thirties, I thought 65 was ancient. Even in my fifties, I thought that by 85 I would be well past it. No doubt many of my readers think likewise. But suppose someone had told me that if I did a few relatively simple things, I could hone my brain to be as sharp as a razor and keep it that way? And that these things were not only simple but almost all of them completely non-onerous, even enjoyable? Furthermore, that I would not be wasting my time on fashionable fads – these measures would all be verified by science? Well, that is where we are now and we are fortunate. *When I was 30, none of this evidence even existed.* Over the past few decades, neuroscience, psychology and psychiatry have between them pulled the cat out of the bag – the secret of supercharging our brain. Some aspects of this secret might sound predictable: eat the right things, don't overindulge in alcohol and get enough sleep. But scientists are picky people; we don't like putting out messages which have no proof to sustain them, 'common sense' though they may be. Now we know for certain that those measures do make a big difference to the brain – and that difference can be quantified.

But there are also entirely new and unexpected stories from science that enrich the picture immeasurably. To mention but a few of those I have discussed in this book: the bugs in your bowel greatly influence your brain's performance – astonishing! Don't just clean your teeth – chew gum, eat apple cores. Really? Yes! And who'd have thought that the frequency and quality of your orgasms will keep your brain going (and need not lessen over time)? The big message, the core message, is this: although there is no single, magic bullet – no pill we can take, no simple ten-day plan, no one remedy – by applying the lessons of science day by day across the course of our lives we shall stay on a winning trajectory. Doesn't genetics make a difference, I hear you say? Yes – but not as much as you may think. Remember the findings of The Disconnected Mind, that wonderful Scottish brain study? About half of our fluid intelligence can be backtracked to our childhood IQ; but of the remaining half, the majority of the predictors of our brain power are under our own control.

Investing in your brain is like investing in the stock market. It's not the moment at which you invest that matters most; it's the time you stay in the market. You may get lucky by gambling, but this is not a recipe I would apply to my health! In terms of brain power, it's not what you do on any single day that counts, but what you do over many years, every day, building up the cumulative effects of your lifestyle. And, like compound financial interest, the gains on offer are prodigious.

This does not mean that we have to slavishly follow a restrictive programme or live a downbeat lifestyle. Just as what we do day by day, most days, matters, what we do on one particular day is not going to kill our brains stone dead. It's not one night on the town that is going to get us. It's if we overdo it every day or even, in some circumstances, every week. Drinking a bottle of wine a day – or even a bottle of vodka – is not unusual for some. I would not recommend it. But celebrating and letting your hair down occasionally is not only unlikely to kill you; it will contribute to improving your quality of life.

You may ask: of all the many messages in this book, are there any 'stand-outs'? I believe there are, and I'll close here with three of them. The biggest of these to me – it is a personal view – would be: exercise. Many people dread hearing this. It is so easy to slide into a sedentary lifestyle and so difficult, with commuting and work and family commitments, to keep notching up the recommended 150 minutes of aerobic exercise every week. But somehow we have to find the time – and, above all, a way to make it enjoyable. We don't have to slog it out at the gym or go running in the rain. Anything to lighten it – do it in a group, walk the dog, choose an activity you like. And even if it's not 150 minutes, *anything* you do to increase your activity levels, over a lifetime, will make a difference.

Second is a more unusual one, considered in several chapters: stay within your daily rhythms. Recently, a national newspaper in the UK ran a health section called 'Your perfect day according to science'. It was all about *not* doing things at the wrong time of day and doing the right things at the right time. Our bodies are virtually a collective bag of rhythms, embedded over millions of years *in our brains*. I would not recommend fighting them! Constantly varying bedtimes, unguarded excesses at all times of the day and night, jet-lag, late-night working and irregular habits of all kinds – across a lifetime, these disruptive patterns will conspire to blunt the brain. Here again, you don't have to rule out the occasional irregularity: the odd eccentricity will not do any permanent damage to your brain performance. But constant abuse will.

And the last of my big three? This would be social capital and social connectivity. Here I'd just remind you of that amazing finding of the Harvard Adult Development Study, discussed in chapter 10: that participants from all backgrounds, rich and poor, all said at the start of their lives that the big thing that mattered was working hard, achieving fame, becoming rich – and yet at the end they had *all* changed their minds. The thing that mattered was friends and family. That made a huge impact on me. Furthermore, it wasn't just their opinion. The objective findings of the Harvard investigators showed that those who had good-quality relationships – friends who cared, family who were dependable – were the ones who had better health and well-being. And, as other evidence clearly shows, better brains in the long term.

For all I have written and you have read, the best is yet to come. The horizon is full of promise. We know a great deal now about how we can supercharge our brains. But this wonderful, complex organ, to which we owe the richness of our human experience, is only just being unravelled. Forget the claims of AI. When a super-server can juggle, fall in love with its dog or feel the depths of Shakespeare's greatest poetry, then I shall believe that it has got somewhere near the average human brain. Around the corner in the coming decades, science will reveal even more secrets of how we can nurture this virtuous organ. Stay sharp, stay sure and stay certain that the best is yet to come.

NOTES

INTRODUCTION

1 The Global Council on Brain Health (GCBH) is an independent collaborative of scientists, health professionals, scholars and policy experts from around the world who are working in areas of brain health related to human cognition. It is based in Washington DC and funded by AARP, the leading organization for people aged over 50 in the USA.

1: MEET YOUR BRAIN

1 '5 unsolved mysteries about the brain' (Seattle: Allen Institute for Brain Science, 14 March 2019), https://alleninstitute.org/what-we-do/brain-science/news-press/articles/5-unsolved-mysteries-about-brain.

2: BODY AND MIND

1 Hilary Hylton, 'Runner's high: joggers live longer', *Time*, 12 Aug. 2008, http://content.time.com/time/health/article/0,8599,1832033,00.html.

2 T. M. Manini, 'Energy expenditure and aging', *Ageing Research Reviews*, Vol. 9, No. 1, 2010, p. 9.

3 Susan McQuillan, 'Fidgeting has benefits', *Psychology Today*, 17 Sept. 2016.

3: GODZILLA BRAIN

1 Giulia Enders, *Gut*, trans. David Shaw (London: Scribe, 2015).

4: BUGS IN THE BRAIN

1 It's easy to calculate your BMI. Measure your height in metres and multiply that number by itself. Now weigh yourself in kilograms. Then divide the

second number by the first. So, for example, if you're 1.75m tall and weigh 70kg, your BMI is 70 divided by 3, which is just over 23 – within the 'healthy range' of 18–25.

2   K. A. Dill-McFarland et al., 'Close social relationships correlate with human gut microbiota composition', *Nature Scientific Reports*, Vol. 9, Article 703, 2019, https://doi.org/10.1038/s41598-018-37298-9.

3   Charles Darwin, *The Expression of the Emotions in Man and Animals* (London: John Murray, 1872).

4   A robust finding in science is one that survives replication: in other words, repeated experiments always produce the same findings.

5   John F. Cryan and Timothy G. Dinan, 'Mind-altering microorganisms: the impact of the gut microbiota on brain and behaviour', *Nature Reviews Neuroscience*, Vol. 13, No. 10, 2012, p. 701.

5: FOOD FOR THOUGHT

1   Suartcha Prueksaritanond et al., 'A puzzle of hemolytic anemia, iron and Vitamin B12 deficiencies in a 52-year-old male', *Case Reports in Hematology*, Vol. 2013, art. ID 708489.

2   Cited in Harri Hemilä, 'A brief history of vitamin C and its deficiency, scurvy', open access paper, Department of Public Health, University of Helsinki, 2006, https://www.mv.helsinki.fi/home/hemila/history/.

3   Anne W. S. Rutjes et al., 'Vitamin and mineral supplementation for maintaining cognitive function in cognitively healthy people in mid and late life', Cochrane Database of Systematic Reviews, 17 Dec. 2018, p. 2, DOI: 10.1002/14651858.CD011906.pub2.

4   Global Council on Brain Health, *The Real Deal on Brain Health Supplements: GCBH recommendations on vitamins, minerals, and other dietary supplements*, 2019, p. 4, https://doi.org/10.26419/pia.00094.001.

6: NO BRAIN IS AN ISLAND

1   B. A. Primack et al., 'Positive and negative experiences on social media and perceived social isolation', *American Journal of Health Promotion*, Vol. 33, No. 6, 2019, pp. 859–68.

2   John Milton, *Paradise Lost*, book II.

3   Jill Lepore, 'A history of loneliness', *New Yorker*, 6 April 2020.

4   Charles Dickens, *American Notes for General Circulation* (London: Chapman and Hall, 1842), cited in S. Gallagher, 'The cruel and unusual phenomenology of solitary confinement', in *Frontiers in Psychology*, Vol. 5, 2014.

5 I. E. M. Evans, A. Martyr, R. Collins, C. Brayne and L. Clare, 'Social isolation and cognitive function in later life: a systematic review and meta-analysis', *Journal of Alzheimer's Disease*, Vol. 70, Suppl. 1, 2019, pp. S119–44.

6 Michael Babula, *Motivation, Altruism, Personality and Social Psychology: the coming age of altruism* (New York: Springer, 2013).

7 R. S. Weiss, *Loneliness: the experience of emotional and social isolation* (Cambridge, MA: MIT Press, 1972).

7: SEX ON THE BRAIN

1 'Understanding the id, ego and superego in psychology', n.d., https://www.dummies.com/education/psychology/understanding-the-id-ego-and-superego-in-psychology/.

2 K. Kapparis, 'Aristophanes, Hippocrates and sex-crazed women', *Ageless Arts: The Journal of the Southern Association for the History of Medicine and Science*, Vol. 1, 2015, pp. 155–70, http://www.sahms.net/uploads/3/4/7/5/34752561/kapparis_final.pdf.

3 Cited in Katherine Harvey, 'The salacious middle ages', 23 Jan. 2018, https://aeon.co/essays/getting-down-and-medieval-the-sex-lives-of-the-middle-ages.

4 Dr Ashton Brown, cited in Therese Oneill, *Unmentionable: The Victorian Lady's Guide to Sex, Marriage and Manners* (London: Little, Brown, 2016).

5 Philip Larkin, 'Annus Mirabilis' (1967).

6 C. Beekman, '1950s discourse on sexuality', 2013, http://social.rollins.edu/wpsites/thirdsight/2013/04/11/1950s-discourse-on-sexuality/.

7 W. H. Masters and V. E. Johnson, *Homosexuality in perspective* (Boston: Little, Brown, 1979), p. 11.

8 'See what science says about women's pleasure', https://www.omgyes.com/.

9 Helen Rumbelow, 'Yes, yes, yes! The app that will turn you on', *The Times*, 1 March 2016.

10 I Corinthians, chapter 7, verse 1.

11 P. Elwood, J. Galante, J. Pickering, S. Palmer, A. Bayer, V. Ben-Shlomo et al., 'Healthy lifestyles reduce the incidence of chronic diseases and dementia: evidence from the Caerphilly Cohort Study', *PLoS ONE*, Vol. 8, No. 12, 2013, e81877.

12 'Benefits of love and sex', https://fisd.oxfordshire.gov.uk/kb5/oxfordshire/directory/advice.page?id=YhqER5vFjpA.

13 R. M. Anderson, 'Positive sexuality and its impact on overall well-being', *Bundesgesundheitsblatt*, Vol. 56, 2013, pp. 208–14, https://doi.org/10.1007/s00103-012-1607-z.

14 S. Brody, 'The relative health benefits of different sexual activities', *Journal of Sexual Medicine*, Vol. 7, No. 4, 2010, pp. 1336–61.

15 Hui Liu et al., 'Is sex good for your health?', *Journal of Health and Social Behaviour*, Vol. 57, No. 3, 2016, pp. 276–96 (emphasis added).

16 *Science Daily*, 22 June 2017, https://www.sciencedaily.com/releases/2017/06/170622083020.htm.

17 James H. Clark, 'A critique of Women's Health Initiative Studies (2002–2006)', *Nuclear Receptor Signaling*, Vol. 4, 2006, e023, DOI: 10.1621/nrs.04023.

18 National Academies of Sciences, Engineering and Medicine, *The Clinical Utility of Compounded Bioidentical Hormone Therapy: a review of safety, effectiveness, and use* (Washington DC, 2020).

19 'Hormone therapy: is it right for you?', Mayo Clinic, 9 June 2020, https://www.mayoclinic.org/diseases-conditions/menopause/in-depth/hormone-therapy/art-20046372.

8: MIND GAMES

1 Leonardo da Vinci, *A Treatise on Painting* (New York: Dover, 2005), unabridged re-issue of John Francis Rigaud's translation as published by George Bell & Sons, London, 1877, pp. 4, 7.

2 K. Rehfeld et al., 'Dance training is superior to repetitive physical exercise in inducing brain plasticity in the elderly', *PLoS ONE*, Vol. 13, No. 7, 2018, e0196636, https://journals.plos.org/plosone/article?id=10.1371/journal.pone.0196636.

3 Ibid.

4 Global Council on Brain Health, *Engage Your Brain: GCBH recommendations on cognitively stimulating activities*, 2017, p. 4, https://www.aarp.org/content/dam/aarp/health/brain_health/2017/07/gcbh-cognitively-stimulating-activities-report.pdf.

5 Stanford Center on Longevity, *A Consensus on the Brain Training Industry from the Scientific Community*, 20 Oct. 2014, http://longevity.stanford.edu/a-consensus-on-the-brain-training-industry-from-the-scientific-community-2/.

6 Cognitive Training Data, *Cognitive Training Data Response Letter*, 2014, https://www.cognitivetrainingdata.org/the-controversy-does-brain-training-work/response-letter/.

7 A. M. Owen et al., 'Putting brain training to the test', *Nature*, Vol. 465, No. 7299, 10 June 2010, p. 778.

8 Global Council on Brain Health, *Engage Your Brain*. Washington DC: AARP.

9: TO SLEEP OR NOT TO SLEEP?

1 M. de Manaceine, 'Quelques observations experimentales sur l'influence de l'insomnie absolue', *Archives Italiennes de Biologie*, Vol. 21, 1894, pp. 322–5.

2 Cited in M. K. Scullin and D. L. Bliwise, 'Sleep, cognition, and normal aging: integrating a half century of multidisciplinary research', *Perspectives on Psychological Science*, Vol. 10, No. 1, 2015, pp. 97–137.

3 Hans Berger, *Psyche* (Jena: Gustav Fischer, 1940), p. 6.

4 Global Council on Brain Health, *The Brain–Sleep Connection: GCBH recommendations on sleep and brain health* (Washington DC: AARP, 2016).

5 'Women have younger brains than men', *Wall Street Journal*, 27 March 2019.

10: THE FEEL-GOOD FACTOR

1 Viktor Frankl, *Man's Search for Meaning* (New York: Washington Square Books, 1984).

2 David J. Llewellyn et al., 'Cognitive function and psychological well-being: findings from a population-based cohort', *Age and Ageing*, Vol. 37, No. 6, 2008, p. 687.

3 David Hume, *A Treatise of Human Nature*, Book III, Part III, Section III, 'Of the influencing motives of the will' (1739).

4 https://www.berkeleywellbeing.com/well-being-survey.html.

5 Stephen Hawking, speech at opening of the Paralympic Games in London, 2012.

6 M. A. Nattali, *The Hindoos*, Vol. 2 (London, 1846), ch. 10.

7 Carolyn Gregoire, 'What India can teach the rest of the world about living well', *Huffington Post*, 11 Nov. 2013.

8 N. P. Gothe, I. Khan, J. Hayes, E. Erlenbach and J. S. Damoiseaux, 'Yoga effects on brain health: a systematic review of the current literature', *Brain Plasticity*, Vol. 5, No. 1, 2019, pp. 105–22.

9 Kaushik Talukdar, 'Breath of life', *Telegraph* online, 8 Dec. 2020, https://www. telegraphindia.com/health/breath-of-life-lets-explore-the-science-of-pranayama/cid/1785301.

10 Semir Zeki, 'Artistic creativity and the brain', *Science*, Vol. 293, No. 5527, 2001, pp. 51–2.

11 http://www.healing-power-of-art.org/.

12 M. A. Sayette, 'Does drinking reduce stress?', *Alcohol Research & Health: The Journal of the National Institute on Alcohol Abuse and Alcoholism*, Vol. 23, No. 4, 1999, pp. 250–5.

13 A. Abbey et al., 'Subjective, social, and physical availability, II: their simultaneous effects on alcohol consumption', *International Journal of the Addictions*, Vol. 25, 1990, pp. 1011–23.

14 Séverine Sabia et al., 'Alcohol consumption and cognitive decline in early old age', *Neurology*, Vol. 82, No. 4, 2014, pp. 332–9.

15 C. A. Paul et al., 'Association of alcohol consumption with brain volume in the Framingham study', *Archives of Neurology*, Vol. 65, No. 10, 1008, pp. 1363–7.

16 Angela M. Wood et al., 'Risk thresholds for alcohol consumption: combined analysis of individual-participant data for 599,912 current drinkers in 83 prospective studies', *Lancet*, Vol. 391, No. 10129, 2018, pp. 1513–23.

11: COVID-19 AND THE BRAIN

1 Cited by Mark Honigsbaum and Lakshmi Krishnan in *Lancet*, Vol. 396, No. 10260, 31 Oct. 2020, pp. 1389–91.

2 Abhinandan Mishray, 'PLA-controlled Wuhan lab under fresh scanner', *Sunday Guardian*, 15 May 2021, https://www.sundayguardianlive.com/news/pla-controlled-wuhan-lab-fresh-scanner.

3 David McKenna, 'Eyam plague: the village of the damned', BBC News, 5 Nov. 2016, https://www.bbc.co.uk/news/uk-england-35064071.

4 H. Bundgaard, J. S. Bundgaard, Daniel Emil Tadeusz Raaschou-Pedersen et al., 'Effectiveness of adding a mask recommendation to other public health measures to prevent SARS-CoV-2 infection in Danish mask wearers – a randomized controlled trial', *Annals of Internal Medicine*, Vol. 174, 2021, pp. 335–43, https://doi.org/10.7326/M20-6817.

5 Andrew G. Letizia, Irene Ramos, Ajay Obla et al., 'SARS-CoV-2 transmission among marine recruits during quarantine', *New England Journal of Medicine*, Vol. 383, No. 25, Dec. 2020, pp. 2407–16, DOI: 10.1056/NEJMoa2029717.

6 Nature News Feature: 'Facemasks: what the data say', news feature, *Nature*, 6 Oct. 2020, https://www.nature.com/articles/d41586-020-02801-8.

7 Lisa M. Brosseau and Margaret Sietsema, 'Masks-for-all for COVID-19 not based on sound data', editorial, Center for Infectious Disease Research and Policy, University of Minnesota, https://www.cidrap.umn.edu/news-perspective/2020/04/commentary-masks-all-covid-19-not-based-sound-data.

8 N. H. L. Leung, D. K. W. Chu, E. Y. C. Shiu et al., 'Respiratory virus shedding in exhaled breath and efficacy of face masks', *Nature Medicine*, Vol. 26, No. 5, 2020, pp. 676–80.

9 Dr Colin Axon, Brunel University, UK, cited in Justin Stoneman, 'Cloth face masks are "comfort blankets" that do little to curb Covid spread',

Telegraph, 17 July 2021, https://www.telegraph.co.uk/news/2021/07/17/cloth-face-masks-comfort-blankets-do-little-curb-covid-spread/.

10 B. G. Iversen, D. F. Vestrheim, S. Flottorp, E. Denison and A. D. Oxman, *Should individuals in the community without respiratory symptoms wear facemasks to reduce the spread of COVID-19? A rapid review* (Oslo: Norwegian Institute of Public Health 2020), https://www.fhi.no/globalassets/dokumenterfiler/rapporter/2020/should-individuals-in-the-community-without-respiratory-symptoms-wear-facemasks-to-reduce-the-spread-of-covid-19-report-2020.pdf.

11 Tom Jefferson and Carl Heneghan, *Masking lack of evidence with politics* (Oxford: Centre for Evidence-Based Medicine, 23 July 2020), https://www.cebm.net/covid-19/masking-lack-of-evidence-with-politics/.

12 M. Taquet, S. Luciano, J. R. Geddes and P. J. Harrison, 'Bidirectional associations between COVID-19 and psychiatric disorder: retrospective cohort studies of 62 354 COVID-19 cases in the USA', *Lancet Psychiatry*, Vol. 8, No. 2, Feb. 2021, pp. 130–40; first publ. online 9 Nov. 2020, https://doi.org/10.1016/S2215-0366(20)30462-4.

13 Office for National Statistics, *Coronavirus and depression in adults, Great Britain: January to March 2021* (London, 2021).

14 The Health Foundation, 'Latest data highlights a growing mental health crisis in the UK', press release, 5 May 2021, https://www.health.org.uk/news-and-comment/news/latest-data-highlights-a-growing-mental-health-crisis-in-the-uk.

15 'How Covid upended a century of patterns in US deaths', *New York Times*, 23 April 2021, https://www.nytimes.com/interactive/2021/04/23/us/covid-19-death-toll.html.

16 G. A. de Erausquin, H. Snyder, M. Carrillo, A. A. Hosseini, T. S. Brugha and S. Seshadri, 'The chronic neuropsychiatric sequelae of COVID-19: the need for a prospective study of viral impact on brain functioning', *Alzheimer's and Dementia*, Vol. 17, No. 6, pp. 1056–65, first publ. online 5 Jan. 2021, https://doi.org/10.1002/alz.12255.

17 Bianca Nogrady, 'What the data say about asymptomatic COVID infections', news report, *Nature*, 18 Nov. 2020.

18 *Transmission of SARS-CoV-2: implications for infection prevention precautions*, scientific brief, (Geneva: WHO, 9 July 2020), https://www.who.int/news-room/commentaries/detail/transmission-of-sars-cov-2-implications-for-infection-prevention-precautions.

19 *SARS-CoV-2 and surface (fomite) transmission for indoor community environments*, CDC Science Brief (Atlanta: Centers for Disease Control and Prevention, 5 April 2021), https://www.cdc.gov/coronavirus/2019-ncov/more/science-and-research/surface-transmission.html.

20 M. Z. Tay, C. M. Poh, L. Rénia et al. 'The trinity of COVID-19: immunity, inflammation and intervention', *Nature Reviews Immunology*, Vol. 20, 2020, pp. 363–74, https://doi.org/10.1038/s41577-020-0311-8.

21 Huijing Xia and Eric Lazartigues, 'Angiotensin-converting enzyme 2 in the brain: properties and future directions', *Journal of Neurochemistry*, Vol. 107, No. 6, 2008, pp. 1482–94, doi:10.1111/j.1471-4159.2008.05723.x.

22 M. Z. Tay, C. Poh, L. Rénia et al., 'The trinity of COVID-19: immunity, inflammation and intervention', *Nature Reviews Immunology*, Vol. 20, 2020, pp. 363–74, https://doi.org/10.1038/s41577-020-0311-8.

23 'COVID-19 and the brain: What do we know so far?', *Medical News Today*, 25 Jan. 2021.

24 Michael Marshall, 'How COVID-19 can damage the brain', news feature, *Nature*, 15 Sept. 2020, https://www.nature.com/articles/d41586-020-02599-5.

25 Katarina Zimmer, 'COVID-19's effects on the brain', *The Scientist*, 20 Jan. 2021, https://www.the-scientist.com/news-opinion/covid-19s-effects-on-the-brain-68369.

26 Michelle A. Erickson, Elizabeth M. Rhea, Rachel C. Knopp and William A. Banks, 'Interactions of SARS-CoV-2 with the blood–brain barrier', *International Journal of Molecular Sciences*, Vol. 22, No. 5, 2021, article 2681, doi:10.3390/ijms22052681.

27 Jocelyn Solis-Moreira, 'Study identifies how SARS-CoV-2 enters and infects the brain', news report, *Medical Life Sciences*, 21 April 2021, https://www.news-medical.net/news/20210421/Study-identifies-how-SARS-CoV-2-enters-and-infects-the-brain.aspx.

28 F. Seehusen, Jordan J. Clark, Parul Sharma et al., 'Viral neuroinvasion and neurotropism without neuronal damage in the hACE2 mouse model of COVID-19', *bioRxiv*, 2021, https://doi.org/10.1101/2021.04.16.440173.

29 R. Costa, E. Burgos-Ramos, O. Gomez Torres and D. Cruz-Topete, 'Multi-organ effects of SARS-CoV2, more than a respiratory virus: effects on human astrocytes', *FASEB Journal*, Vol. 35, Suppl. 1, 2021, https://doi.org/10.1096/fasebj.2021.35.S1.02430.

30 W. A. Banks, 'Passage of cytokines across the blood–brain barrier', *Neuroimmunomodulation*, Vol. 2, No. 4, 1995, pp. 241–8.

31 Erickson et al., 'Interactions'.

32 P. Kumari, H. A. Rothan, J. P. Natekar et al., 'Neuroinvasion and encephalitis following intranasal inoculation of SARS-CoV-2 in K18-hACE2 mice', *Viruses*, Vol. 13, no. 1, 2021, p. 132, doi:10.3390/v13010132.

33 J. Meinhardt, J. Radke, C. Dittmayer et al., 'Olfactory transmucosal SARS-CoV-2 invasion as a port of central nervous system entry in individuals with

COVID-19', *Nature Neuroscience*, Vol. 24, 2021, pp. 168–75, first publ. online 30 Nov. 2020, doi: 10.1038/s41593-020-00758-5.

34 Erickson et al., 'Interactions'.

35 W. Trypsteen, J. Van Cleemput, W. Snippenberg et al., 'On the whereabouts of SARS-CoV-2 in the human body: a systematic review', *PLoS Pathogens*, Vol. 16, No. 10, 2020, e100903.

36 M. Honigsbaum and L. Krishnan, 'Taking pandemic sequelae seriously: from the Russian influenza to COVID-19 long-haulers', *Lancet*, Vol, 396, No. 10260, 31 Oct. 2020, pp. 1389–91, https://doi.org/10.1016/S0140-6736(20)32134-6.

37 Andrew Woodcock, 'Thousands facing long Covid as "collateral damage" of Boris Johnson's decision to lift restrictions', *Independent*, 6 July 2021, https://www.independent.co.uk/news/uk/politics/long-covid-boris-johnson-whitty-b1879275.html.

38 Edith L. Graham, Jeffrey R. Clark, Zachary S. Orban et al., 'Persistent neurologic symptoms and cognitive dysfunction in non-hospitalized Covid-19 "long haulers"', *Annals of Clinical and Translational Neurology*, Vol. 8, No. 5, 2021, pp. 1073–85.

39 Natalia de las Heras, Virna Margarita Martin Giménez, León Ferder et al., 'Implications of oxidative stress and potential role of mitochondrial dysfunction in COVID-19' (2020). *Antioxidants*, 9, 9, 897. Available at: https://doi.org/10.3390/antiox9090897.

40 G. B. Stefano, R. Ptacek, H. Ptackova, A Martin and R. M. Kream, 'Selective neuronal mitochondrial targeting in SARS-CoV-2 infection affects cognitive processes to induce "brain fog" and results in behavioral changes that favor viral survival', editorial, *Medical Science Monitor*, Vol. 27, 25 Jan. 2021, e930886, https://doi.org/10.12659/MSM.930886.

41 G. L. Nicolson, 'Mitochondrial dysfunction and chronic disease: treatment with natural supplements', *Integrative Medicine* (Encinitas, CA), Vol. 13, No. 4, 2014, pp. 35–43.

42 A. Hampshire, W. Trender, S. R. Chamberlain et al., 'Cognitive deficits in people who have recovered from COVID-19 relative to controls: an N=84,285 online study', *medrxiv*, 2020, https://www.medrxiv.org/content/10.1101/2020.10.20.20215863v1.article-metrics.

43 C. Iddon, A. Hathaway and S. Fitzgerald, 'CIBSE COVID-19 ventilation guidance', *CIBSE Journal*, Aug. 2020.

44 1 μm = one-millionth of a metre or one-thousandth of a millimetre (one micron). One metre = 39 inches. 1 μm = 0.00004 inches.

45 Ming Hui Chua, Weiren Cheng, Shermin Simin Goh et al., 'Face masks in the new COVID-19 normal: materials, testing, and perspectives', *PubMed Research*, 7 Aug. 2020, article 7286735, doi:10.34133/2020/7286735.

46 R. Mittal, R. Ni and J-.H. Seo, 'The flow physics of COVID-19', *Journal of Fluid Mechanics*, Vol. 894, 2020, F2, doi:10.1017/jfm.2020.330.

47 Ibid.

48 W. H. Seto, D. Tsang, R. W. Yung et al., 'Effectiveness of precautions against droplets and contact in prevention of nosocomial transmission of severe acute respiratory syndrome (SARS)', *Lancet*, Vol. 361, No. 9368. 3 May 2003, pp. 1519–20.

49 R. B. Patel, S. D. Skaria, M. M. Mansour and G. C. Smaldone, 'Respiratory source control using a surgical mask: an in vitro study', *Journal of Occupational and Environmental Hygiene*, Vol. 13, No. 7, 2016, pp. 569–76.

50 Jennifer N. Morey, Ian A. Boggero, April B. Scott and Suzanne C. Segerstrom, 'Current directions in stress and human immune function', *Current Opinion in Psychology*, Vol. 5, 2015, pp. 13–17, doi:10.1016/j.copsyc.2015.03.007.

51 P. Haake, T. H. Krueger, M. U. Goebel et al., 'Effects of sexual arousal on lymphocyte subset circulation and cytokine production in man', *Neuroimmunomodulation*, Vol. 11, No. 5, 2004, pp. 293–8, doi: 10.1159/000079409.

52 W. A. Banks, M. J. Reed, A. F. Logsdon et al., 'Healthy aging and the blood–brain barrier', *Nature Aging*, Vol. 1, 2021, pp. 243–54, https://doi.org/10.1038/s43587-021-00043-5.

53 G. A. Rossi, O. Sacco, A. Capizzi and P. Mastromarino, 'Can resveratrol-inhaled formulations be considered potential adjunct treatments for COVID-19?', *Frontiers in Immunology*, 19 May 2021, https://doi.org/10.3389/fimmu.2021.670955.

54 C. Toda, J. D. Kim, D. Impellizzeri et al., 'UCP2 regulates mitochondrial fission and ventromedial nucleus control of glucose responsiveness', *Cell*, Vol. 164, No. 5, 2016, pp. 872–83.

55 P. S. Hafen, N. P. Coray, J. R. Sorensen et al., 'Repeated exposure to heat stress induces mitochondrial adaptation in human skeletal muscle', *Journal of Applied Physiology*, Vol. 125, No. 5, 2018, pp. 1447–55.

56 M. Picard et al., 'An energetic view of stress: focus on mitochondria', *Frontiers in Neuroendocrinology*, Vol. 49, 2018, pp. 72–85.

57 Snezana Agatonovic-Kustrin, Ella Kustrin and David W. Morton, 'Essential oils and functional herbs for healthy aging', *Neural Regeneration Research*, Vol. 14, No. 3, 2019, pp. 441–5, doi:10.4103/1673-5374.245467.

58 D. S. Tshibangu, S. Matondo and E. M. Lengbiye, 'Possible effect of aromatic plants and essential oils against COVID-19: review of their antiviral activity', *Journal of Complementary and Alternative Medical Research*, Vol. 11, No. 1, 2020, pp. 10–22.

59 H. Javed, M. Meeran, N. K. Jha and S. Ojha, 'Carvacrol, a plant metabolite targeting viral protease (M^{pro}) and ACE2 in host cells can be a possible candidate for COVID-19', *Frontiers in Plant Science*, Vol. 11, 2021, article 601335, https://doi.org/10.3389/fpls.2020.601335.

SOURCES

INTRODUCTION

Frank L. Baum (1900). *The Wonderful Wizard of Oz*. Chicago: George M. Hill.

Martyn Harlow (1848). 'Passage of an iron rod through the head', *Boston Medical and Surgical Journal* 39 (20), 389–93. Boston: David Clapp.

I. J. Deary, L. J. Whalley and J. M. Starr (2009). *A Lifetime of Intelligence: follow-up studies of the Scottish mental surveys of 1932 and 1947*. Washington DC: American Psychological Association.

Séverine Sabia et al. (2018). 'Alcohol consumption and risk of dementia: 23 year follow-up of Whitehall II cohort study', *British Medical Journal*, 362, k2927.

1: MEET YOUR BRAIN

Thomas S. Kuhn (1962). *The Structure of Scientific Revolutions*. Chicago: University of Chicago Press.

Madhura Ingalhalikar et al. (2014). 'Sex differences in the structural connectome of the human brain', *Proceedings of the National Academy of Sciences of the United States of America*, 111 (2), 823–8.

Richard Nisbett et al. (2012). 'Intelligence: new findings and theoretical developments', *American Psychologist* 67 (2), 130–59.

Arthur R. Jensen (1969). 'How much can we boost IQ and scholastic achievement?', *Harvard Educational Review* 39 (1), 1–123.

C. G. Phillips (1973). Hughlings Jackson Lecture: 'Cortical localization and "sensori motor processes" at the "middle level" in primates', *Proceedings of the Royal Society of Medicine* 66 (10), 987–1002.

C. Daniel Salzman (2011). 'The neuroscience of decision making', *Kavli Foundation Newsletter*, Aug.

Paul E. Dux et al. (2009). 'Training improves multitasking performance by increasing the speed of information processing in human prefrontal cortex', *Neuron* 63 (1), 127–38.

Daniel W. Belsky (2015). 'Quantification of biological aging in young adults', *Proceedings of the National Academy of Sciences of the United States of America* 112 (30), E4104–10.

Norman Doidge (2007). *The Brain that Changes Itself*. New York: Viking.

E. P. Vining (1997). 'Why would you remove half a brain? The outcome of 58 children after hemispherectomy – the Johns Hopkins experience: 1968 to 1996', *Pediatrics*, 100 (2 Pt 1), 163–71.

T. Ngandu (2015). 'A 2 year multidomain intervention of diet, exercise, cognitive training, and vascular risk monitoring versus control to prevent cognitive decline in at-risk elderly people (FINGER): a randomised controlled trial', *Lancet* 385 (9984), P2255–63.

Ian J. Deary (2012). 'Genetic contributions to stability and change in intelligence from childhood to old age', *Nature* 482 (7384), 212–15.

J. G. Makin, D. A. Moses and E. F. Chang (2020). 'Machine translation of cortical activity to text with an encoder–decoder framework', *Nature Neuroscience* 23 (4), 575–82.

E. P. Moreno-Jiménez, M. Flor-García, J. Terreros-Roncal et al. (2019). 'Adult hippocampal neurogenesis is abundant in neurologically healthy subjects and drops sharply in patients with Alzheimer's disease', *Nature Medicine* 25 (4), 554–60.

D. Hakim, Jason Chami and Kevin A. Keay (2020). 'μ-opioid and dopamine-D2 receptor expression in the nucleus accumbens of male Sprague-Dawley rats whose sucrose consumption, but not preference, decreases after nerve injury', *Behavioural Brain Research* 381, art. no. 112416.

S. Kim et al. (2019). 'Transneuronal propagation of pathologic α-synuclein from the gut to the brain models Parkinson's disease', *Neuron* 103 (4), pp. 627–41, e7.

Lou Beaulieu-Laroche (2018). 'Enhanced dendritic compartmentalization in human cortical neurons', *Cell* 175 (3), p. 643.

S. Reardon (2019). 'Pig brains kept alive outside body for hours after death', *Nature* 568, pp. 283–4.

S. Reardon (2020). 'Can lab-grown brains become conscious?', *Nature* 586, pp. 658–61.

2: BODY AND MIND

Michel Poulain and Giovanni Mario Pes (2004). 'Identification of a geographic area characterized by extreme longevity in the Sardinia island: the AKEA study', *Experimental Gerontology* 39 (9), 1423–9.

Dan Buettner (2009). *The Blue Zones: lessons for living longer from the people who've lived the longest*. Washington DC: National Geographic.

Charles Lyell (1830–3). *The Principles of Geology*. London: John Murray.

Freeletics Research (2019). *UK research finds majority fear their unhealthy lifestyles will lead to an early grave*. Munich: Freeletics, 10 Jan., 2019.

James Fuller Fixx (1977). *The Complete Book of Running*. New York: Random House.

Kenneth H. Cooper (1985). *Running Without Fear: how to reduce the risk of heart attack and sudden death during aerobic exercise*. New York: Evans.

Eliza F. Chakravarty et al. (2008). 'Reduced disability and mortality among aging runners: a 21-year longitudinal study', *Archives of Internal Medicine* 168 (15), 1638–46.

Frank W. Booth, Christian K. Roberts and Matthew J. Laye (2012). 'Lack of exercise is a major cause of chronic diseases', *Comparative Physiology* 2 (2), 1143–1211.

Marc R. Hamilton et al. (2008). 'Too little exercise and too much sitting: inactivity physiology and the need for new recommendations on sedentary behavior', *Current Cardiovascular Risk Reports* 2 (4), 292–8.

A. H. Shadyab (2017). 'Associations of accelerometer-measured and self-reported sedentary time with leukocyte telomere length in older women', *American Journal of Epidemiology* 185 (3), 172–4.

N. Genevieve et al. (2008). 'Television time and continuous metabolic risk in physically active adults', *Medicine and Science in Sports and Exercise* 40 (4), 639–45.

Ryan S. Falck et al. (2016). 'What is the association between sedentary behaviour and cognitive function? A systematic review', *British Journal of Sports Medicine* 51 (10), 800–11.

Yan Shijiao et al. (2020). 'Association between sedentary behavior and the risk of dementia: a systematic review and meta-analysis', *Translational Psychiatry* 10 (1), 112.

Waneen Wyrick Spirduso (1975). 'Reaction and movement time as a function of age and physical activity level', *Journal of Gerontology* 30 (4), 435–40.

Arthur F. Kramer and Kirk I. Erickson (2007). 'Effects of physical activity on cognition, well-being, and brain: human interventions', *Alzheimer's & Dementia: The Journal of the Alzheimer's Association* 3 (25), S45–51.

Kirk I. Erickson et al. (2011). 'Exercise training increases size of hippocampus and improves memory', *Proceedings of the National Academy of Sciences of the United States of America* 108 (7), 3017–22.

T. M. Manini (2010). 'Energy expenditure and aging', *Ageing Research Reviews* 9 (1), 1–11.

Carl W. Cotman and Nicole C. Berchtold (2002). 'Exercise: a behavioral intervention to enhance brain health and plasticity', *Trends in Neurosciences* 25 (6), 295–301.

David A. Raichlen and Gene E. Alexander (2017). 'Adaptive capacity: an evolutionary-neuroscience model linking exercise, cognition, and brain health', *Trends in Neurosciences* 40 (7), 408–21.

Michael J. Wheeler et al. (2017). 'Sedentary behavior as a risk factor for cognitive decline? A focus on the influence of glycemic control in brain health', *Alzheimer's and Dementia: Translational Research and Clinical Interventions*, 3 (3), 291–300.

Christopher Mark Spray et al. (2006). 'Understanding motivation in sport: an experimental test of achievement goal and self determination theories', *European Journal of Sport Science* 6 (1), 43–51.

J. A. Levine, L. M. Lanningham-Foster, S. K. McCrady et al. (2005). 'Interindividual variation in posture allocation: possible role in human obesity', *Science* 307 (5709), 584–6.

Guohua Zheng (2019). 'Effect of aerobic exercise on inflammatory markers in healthy middle-aged and older adults: a systematic review and meta-analysis of randomized controlled trials', *Frontiers in Aging Neuroscience*, https://doi.org/10.3389/fnagi.2019.00098.

P. Siddarth, A. C. Burggren, H. A. Eyre, G. W. Small and D. A. Merrill (2018). 'Sedentary behavior associated with reduced medial temporal lobe thickness in middle-aged and older adults', *PLoS ONE* 13 (4), e0195549.

Gabriel A. Koepp, Graham K. Moore and James A. Levine (2016). 'Chair-based fidgeting and energy expenditure', *BMJ Open Sport & Exercise Medicine* 2 (1), e000152.

3: GODZILLA BRAIN

Rachel N. Carmody et al. (2016). 'Genetic evidence of human adaptation to a cooked diet', *Genome Biology and Evolution* 8 (4), 1091–1103.

Bruno Bonaz, Thomas Bazin and Sonia Pellissier (2018). 'The vagus nerve at the interface of the microbiota–gut–brain axis', *Frontiers in Neuroscience* 12, 49.

Natasha Bray (2019). 'The microbiota–gut–brain axis', *Nature Research Milestones* 18, S22, June.

William Beaumont (1825). 'A case of wounded stomach', *Philadelphia Medical Recorder*, Jan. (cited in William Beaumont, *Experiments and Observations on the Gastric Juice and the Physiology of Digestion*, New York: Dover, 1959).

Giulia Enders (2015). *Gut*, trans. David Shaw. London: Scribe.

W. O. Atwater (1887). 'The potential energy of food. The chemistry of food. III'. *Century* 34, 397–405 (cited in James L. Hargrove, 'History of the calorie in nutrition', *Journal of Nutrition* 136, 2006, 2957–61).

Leah M. Kalm and Richard D. Semba (2005). 'They starved so that others be better fed: remembering Ancel Keys and the Minnesota experiment', *Journal of Nutrition* 135 (6), 1347–52.

C. M. McCay and Mary F. Crowell (1934). 'Prolonging the life span', *Scientific Monthly* 39 (5), 405–14.

Jasper Most, Valeria Tosti, Leanne M. Redman and Luigi Fontana (2017). 'Calorie restriction in humans: an update', *Ageing Research Reviews* 39, 36–45.

Shin-Hae Lee and Kyung-Jin Min (2013). 'Caloric restriction and its mimetics', *BMB Reports* 46 (4), 181–7.

Yonas E. Geda et al. (2013). 'Caloric intake, aging, and mild cognitive impairment: a population-based study', *Journal of Alzheimer's Disease* 34 (2), 501–7.

A. V. Witte et al. (2009). 'Caloric restriction improves memory in elderly humans', *Proceedings of the National Academy of Sciences of the United States of America* 106 (4), 1255–60.

Jason Brandt et al. (2019). 'Preliminary report on the feasibility and efficacy of the modified Atkins diet for treatment of mild cognitive impairment and early Alzheimer's disease', *Journal of Alzheimer's Disease* 68 (3), 969–81.

Rafael de Cabo and Mark P. Mattson (2019). 'Effects of intermittent fasting on health, aging, and disease', *New England Journal of Medicine* 381 (26), 2541–51.

Mark P. Mattson (2019). 'An evolutionary perspective on why food overconsumption impairs cognition', *Trends in Cognitive Science* 23 (3), 200–12.

4: BUGS IN OUR BRAIN

T. Z. T. Jensen, J. Niemann, K. H. Iversen et al. (2019). 'A 5700-year-old human genome and oral microbiome from chewed birch pitch', *Nature Communications* 10, art. no. 5520.

R. Sender, S. Fuchs and R. Milo (2016). 'Revised estimates for the number of human and bacteria cells in the body', *PLoS Biology* 14 (8), e1002533.

A. Almeida, A. L. Mitchell, M. Boland et al. (2019). 'A new genomic blueprint of the human gut microbiota', *Nature* 568 (7753), 499–504.

D. Zeevi, T. Korem, A. Godneva et al. (2019). 'Structural variation in the gut microbiome associates with host health', *Nature* 568 (7750), 43–8.

G. Falony, M. Joossens, S. Vieira-Silva et al. (2016). 'Population-level analysis of gut microbiome variation', *Science* 352 (6285), 560–4.

Courtney C. Murdock et al. (2017). 'Immunity, host physiology, and behaviour in infected vectors', *Current Opinion in Insect Science* 20, 28–33.

L. Maier et al. (2018). 'Extensive impact of non-antibiotic drugs on human gut bacteria', *Nature* 555 (7698), 623–8.

C. Bressa et al. (2017). 'Differences in gut microbiota profile between women with active lifestyle and sedentary women', *PLoS ONE* 12 (2), e0171352.

K. A. Dill-McFarland, Z. Tang, J. H. Kemis et al. (2019). 'Close social relationships correlate with human gut microbiota composition', *Nature Scientific Reports* 9 (1), 703.

Emily R. Davenport et al. (2014). 'Seasonal variation in human gut microbiome composition', *PLoS ONE* 9 (3), e90731.

Ettje F. Tigchelaar et al. (2015). 'Cohort profile: LifeLines DEEP, a prospective, general population cohort study in the northern Netherlands: study design and baseline characteristics', *BMJ Open* 5 (8), e006772.

M. T. Bailey, S. E. Dowd, J. D. Galley et al. (2011). 'Exposure to a social stressor alters the structure of the intestinal microbiota: implications for stressor-induced immunomodulation', *Brain, Behavior, and Immunity* 25 (3), 397–407.

Siri Carpenter (2012). 'That gut feeling', *Monitor on Psychology* (American Psychological Association) 43 (8), 50.

Birgit Wassermann, Henry Müller and Gabriele Berg (2019). 'An apple a day: which bacteria do we eat with organic and conventional apples?' *Frontiers in Microbiology*, 24 July, https://doi.org/10.3389/fmicb.2019.01629.

M. Schneeberger, A. Everard, A. Gómez-Valadés et al. (2015). 'Akkermansia muciniphila inversely correlates with the onset of inflammation, altered adipose tissue metabolism and metabolic disorders during obesity in mice', *Nature Scientific Reports* 5, 16643.

Robert Caesar et al. (2015). 'Crosstalk between gut microbiota and dietary lipids aggravates WAT inflammation through TLR signaling', *Cell Metabolism* 22 (4), 658–68.

M. Lyte, J. J. Varcoe and M. T. Bailey (1998). 'Anxiogenic effect of subclinical bacterial infection in mice in the absence of overt immune activation', *Physiology and Behaviour* 65 (1), 63–8.

Javier A. Bravo et al. (2011). 'Ingestion of Lactobacillus strain regulates emotional behavior and central GABA receptor expression in a mouse via the vagus nerve', *Proceedings of the National Academy of Sciences of the United States of America* 108 (38), 16050–5.

Kirsten Tillisch et al. (2017). 'Brain structure and response to emotional stimuli as related to gut microbial profiles in healthy women', *Psychosomatic Medicine* 79 (8), 905–13.

Lisa Manderino et al. (2017). 'Preliminary evidence for an association between the composition of the gut microbiome and cognitive function in neurologically healthy older adults', *Journal of the International Neuropsychological Society* 23 (8), 700–5.

A. J. Jeroen (2006). 'Serotonin and human cognitive performance', *Current Pharmaceutical Design* 12 (20), 2473–86.

A. Emeran (2014). 'Gut microbes and the brain: paradigm shift in neuroscience', *Journal of Neuroscience* 34 (46), 15490–6.

N. M. Vogt et al. (2017). 'Gut microbiome alterations in Alzheimer's disease', *Scientific Reports* 7 (1), 13537.

M. Minter, C. Zhang, V. Leone et al. (2016). 'Antibiotic-induced perturbations in gut microbial diversity influences neuro-inflammation and amyloidosis in a murine model of Alzheimer's disease', *Nature Science Reports* 6: 1, 30028.

Stephen S. Dominy et al. (2019). 'Porphyromonas gingivalis in Alzheimer's disease brains: evidence for disease causation and treatment with small-molecule inhibitors', *Science Advances* 5 (1), eaau3333.

Lucy Moss, Andrew Scholey and Keith A. Wesnes (2002). 'Chewing gum selectively improves aspects of memory in healthy volunteers', *Appetite* 38 (3), 235–6.

C. S. Lin, H. H. Lin, S. W. Fann et al. (2020). 'Association between tooth loss and gray matter volume in cognitive impairment', *Brain Imaging and Behavior* 14, 396–407.

John F. Cryan and Timothy G. Dinan (2012). 'Mind-altering microorganisms: the impact of the gut microbiota on brain and behaviour', *Nature Reviews Neuroscience* 13 (10), 701–12.

Elaine Y. Hsiao, Sara W. McBride, Janet Chow, Sarkis K. Mazmanian and Paul H. Patterson (2012). 'Modeling an autism risk factor in mice leads to permanent immune dysregulation', *Proceedings of the National Academy of Sciences of the United States of America* 109 (31), 12776–81.

Filip Scheperjans MD, PhD et al. (2014). 'Gut microbiota are related to Parkinson's disease and clinical phenotype', *Movement Disorders* 30 (3), 350–8.

5: FOOD FOR THOUGHT

Ming-Yi Chiang, Dinah Misner and Gerd Kempermann (1998). 'An essential role for retinoid receptors RARβ and RXRγ in long-term potentiation and depression', *Neuron* 21 (6), P1353–61.

E. Bonnet et al. (2008). 'Retinoic acid restores adult hippocampal neurogenesis and reverses spatial memory deficit in vitamin A deprived rats', *PLoS ONE* 3 (10), e3487.

Coreyann Poly et al. (2011). 'The relation of dietary choline to cognitive performance and white-matter hyperintensity in the Framingham Offspring Cohort', *American Journal of Clinical Nutrition* 94 (6), 1584–91.

Ramon Velazquez et al. (2019). 'Maternal choline supplementation ameliorates Alzheimer's disease pathology by reducing brain homocysteine levels across multiple generations', *Molecular Psychiatry* 25 (10), 2620–9.

Tomasz Huc et al. (2018). 'Chronic low-dose TMAO treatment reduces diastolic dysfunction and heart fibrosis in hypertensive rats', *American Journal of Physiology – Heart and Circulatory Physiology* 315 (6), H1805–20.

Anika K. Smith, Alex R. Wade, Kirsty E. H. Penkman et al. (2017). 'Dietary modulation of cortical excitation and inhibition', *Journal of Psychopharmacology* 31 (5), 632–7.

J. F. Pearson, J. M. Pullar, R. Wilson et al. (2017). 'Vitamin C status correlates with markers of metabolic and cognitive health in 50-year-olds: findings of the CHALICE cohort study', *Nutrients* 9 (8), 831.

Nikolaj Travica et al. (2017). 'Vitamin C status and cognitive function: a systematic review', *Nutrients* 9 (9), 960.

Ibrar Anjum et al. (2018). 'The role of Vitamin D in brain health: a mini literature review', *Cureus* 10 (7), e2960.

David J. Llewellyn et al. (2009). 'Serum 25-hydroxyvitamin D concentration and cognitive impairment', *Journal of Geriatric Psychiatry and Neurology* 22 (3), 188–95.

D. M. Lee et al. (EMAS study group) (2009). 'Association between 25-hydroxyvitamin D levels and cognitive performance in middle-aged and older European men', *Journal of Neurology, Neurosurgery, and Psychiatry* 80 (7), 722–9.

E. Romagnoli et al. (2008). 'Short and long-term variations in serum calciotropic hormones after a single very large dose of ergocalciferol (vitamin D2) or cholecalciferol (vitamin D3) in the elderly', *Journal of Clinical Endocrinology and Metabolism* 93 (8), 3015–20.

M. C. Morris et al. (2002). 'Dietary intake of antioxidant nutrients and the risk of incident Alzheimer disease in a biracial community study', *Journal of the American Medical Association* 287 (24), 3230–7.

F. Mangialasche et al. (2013). 'Serum levels of vitamin E forms and risk of cognitive impairment in a Finnish cohort of older adults', *Experimental Gerontology* 48 (12), 1428–35.

Sahar Tamadon-Nejad et al. (2018). 'Vitamin K deficiency induced by warfarin is associated with cognitive and behavioral perturbations, and alterations in brain sphingolipids in rats', *Frontiers in Aging Neuroscience* 10, 213.

Ludovico Alisi et al. (2019). 'The relationships between Vitamin K and cognition: a review of current evidence', *Frontiers in Neurology* 10, 239.

Inna Slutsky et al. (2010). 'Enhancement of learning and memory by elevating brain magnesium', *Neuron* 65 (2), 165–77.

Enhui Pan et al. (2011). 'Vesicular zinc promotes presynaptic and inhibits postsynaptic long-term potentiation of mossy fiber-CA3 synapse', *Neuron* 71 (6), 1116.

Nicole T. Watt et al. (2010). 'The role of zinc in Alzheimer's disease', *International Journal of Alzheimer's Disease* 2011, 971021.

Aline Thomas et al. (2020). 'Blood polyunsaturated omega-3 fatty acids, brain atrophy, cognitive decline, and dementia risk', *Alzheimer's and Dementia*, Oct., https://doi.org/10.1002/alz.12195.

A. N. Panche, A. D. Diwan and S. R. Chandra (2016). 'Flavonoids: an overview', *Journal of Nutritional Science* 5, e47.

W. T. Wittbrodt and M. Millard-Stafford (2018). 'Dehydration impairs cognitive performance: a meta-analysis', *Medicine & Science in Sports & Exercise* 50, 2360–8.

Ann C. Grandjean and Nicole R. Grandjean (2007). 'Dehydration and cognitive performance', *Journal of the American College of Nutrition* 26 (Suppl. 5), 549S–54S.

Na Zhang, Song M. Du, Jian F. Zhang and Guan S. Ma (2019). 'Effects of dehydration and rehydration on cognitive performance and mood among male college students in Cangzhou, China: a self-controlled trial', *International Journal of Environmental Research and Public Health* 16 (11), 1891.

Rosa Mistica, Coles Ignacio, K-B. Jook and J. Lee (2012). 'Clinical effect and mechanism of alkaline reduced water', *Journal of Food and Drug Analysis* 20 (1), 394–7.

Sanetaka Shirahata, Takeki Hamasaki and Kiichiro Teruya (2012). 'Advanced research on the health benefit of reduced water', *Trends in Food Science & Technology* 23 (2), 124–31.

A. C. van den Brink, E. M. Brouwer-Brolsma, A. A. M. Berendsen and O. van de Rest (2019). 'The Mediterranean, Dietary Approaches to Stop Hypertension (DASH), and Mediterranean-DASH Intervention for Neurodegenerative Delay (MIND) diets are associated with less cognitive decline and a lower risk of Alzheimer's disease – a review', *Advances in Nutrition* 10 (6), 1040–65.

C. T. McEvoy, H. Guyer, K. M. Langa and K. Yaffe (2017). 'Neuroprotective diets are associated with better cognitive function: the Health and Retirement Study', *Journal of the American Geriatrics Society* 65 (8), 1857–62.

Anne W. S. Rutjes et al. (2018). 'Vitamin and mineral supplementation for preventing cognitive deterioration in cognitively healthy people in mid and late life', *Cochrane Database of Systematic Reviews*, 12, art. no.: CD011906.

Dagfinn Aune et al (2017). Fruit and vegetable intake and the risk of cardiovascular disease, total cancer and all-cause mortality – a systematic review and dose-response meta-analysis of prospective studies. *International Journal of Epidemiology*, 46 (3), 1029–1056.

6: NO BRAIN IS AN ISLAND

Igor Borisovich et al. (2014). 'Main findings of psychophysiological studies in the Mars 500 experiment', *Herald of the Russian Academy of Sciences* 84 (2), 106–14.

Esther Herrmann et al. (2007). 'Humans have evolved specialized skills of social cognition: the cultural intelligence hypothesis', *Science* 317 (5843), 1360–6.

John T. Cacioppo, Stephanie Cacioppo and Dorret I. Boomsma (2014). 'Evolutionary mechanisms for loneliness', *Cognition and Emotion* 28 (1), 1–22.

Office of National Statistics (2017). *Loneliness – what characteristics and circumstances are associated with feeling lonely? Analysis of characteristics and circumstances associated with loneliness in England using the Community Life Survey, 2016 to 2017.* London: Office for National Statistics.

Louise C. Hawkley, Kristen Wroblewski, Till Kaiser, Maike Luhmann and L. Philip Schumm (2019). 'Are US older adults getting lonelier? Age, period, and cohort differences', *Psychology and Aging*, 34 (8), 1144–57.

Kali H. Trzesniewski and M. Brent Donnellan (2020). 'Rethinking "Generation Me": a study of cohort effects from 1976–2006', *Perspectives on Psychological Science* 5: 1, 58–75.

D. Matthew, T. Clark, Natalie J. Loxton and Stephanie J. Tobin (2014). 'Declining loneliness over time: evidence from American colleges and high schools', *Personality and Social Psychology Bulletin* 41: 1, 78–89.

B. A. Primack et al. (2019). 'Positive and negative experiences on social media and perceived social isolation', *American Journal of Health Promotion* 33 (6), 859–68.

E. Caitlin, M. S. Coyle and Elizabeth Dugan (2012). 'Social isolation, loneliness and health among older adults', *Journal of Aging and Health* 24 (8), 1346–63.

Gretchen L. Hermes et al. (2009). 'Social isolation dysregulates endocrine and behavioral stress while increasing malignant burden of spontaneous mammary tumors', *Proceedings of the National Academies of Science of the United States of America* 106 (52), 22393–8.

M. Pantell, D. Rehkopf, D. Jutte, S. L. Syme, J. Balmes and N. Adler (2013). 'Social isolation: a predictor of mortality comparable to traditional clinical risk factors', *American Journal of Public Health* 103, 2056–62, doi: 10.2105/AJPH.2013.301261.

J. Holt-Lunstad, T. B. Smith, M. Baker, T. Harris and D. Stephenson (2015). 'Loneliness and social isolation as risk factors for mortality: a meta-analytic review', *Perspectives on Psychological Science* 10 (2), 227–37.

F. R. Day, K. K. Ong and J. R. B. Perry (2018). 'Elucidating the genetic basis of social interaction and isolation', *Nature Communications* 9, art. no. 2457.

Claire Yang et al. (2013). 'Social isolation and adult mortality: the role of chronic inflammation and sex differences', *Journal of Health and Social Behavior* 54 (2), 183–203.

M. Zelikowsky, M. Hui, T. Karigo et al. (2018). 'The neuropeptide Tac2 controls a distributed brain state induced by chronic social isolation stress', *Cell* 173 (5), 1265–79.

G. A. Matthews, E. H. Nieh, C. M. Vander Weele et al. (2016). 'Dorsal raphe dopamine neurons represent the experience of social isolation', *Cell* 164 (4), 617–31.

D. Sargin, D. K. Oliver and E. K. Lambe (2016). 'Chronic social isolation reduces 5-HT neuronal activity via upregulated SK3 calcium-activated potassium channels', *eLife*, e21416, doi:10.7554/eLife.21416.

S. Düzel, J. Drewelies, D. Gerstorf et al. (2019). 'Structural brain correlates of loneliness among older adults', *Nature Science Reports* 9, 13569.

R. Kanai, B. Bahrami, B. Duchaine, A. Janik, M. J. Banissy and G. Rees (2012). 'Brain structure links loneliness to social perception', *Current Biology* 22 (20), 1975–9.

I. E. M. Evans, A. Martyr, R. Collins, C. Brayne and L. Clare (2019). 'Social isolation and cognitive function in later life: a systematic review and meta-analysis', *Journal of Alzheimer's Disease* 70 (S1), S119–44.

V. Heng, M. J. Zigmond and R. J. Smeyne (2018). 'Neurological effects of moving from an enriched environment to social isolation in adult mice', Society for Neuroscience Meeting, San Diego, 2018.

J. Bick, T. Zhu, C. Stamoulis, N. A. Fox, C. Zeanah and C. A. Nelson (2015). 'Effect of early institutionalization and foster care on long-term white matter development: a randomized clinical trial', *JAMA Pediatrics* 169 (3), 211–19.

M. Lehmann, T. Weigel, A. Elkahloun et al. (2017). 'Chronic social defeat reduces myelination in the mouse medial prefrontal cortex', *Nature Science Reports* 7, 46548.

Erin York and Linda Waites (2009). 'Social disconnectedness, perceived isolation, and health among older adults', *Journal of Health and Social Behavior* 50 (1), 31–48.

N. J. Donovan et al. (2017). 'Loneliness, depression and cognitive function in older US adults', *International Journal of Geriatric Psychiatry* 32 (5), 564–73.

E. Lara et al. (2019). 'Does loneliness contribute to mild cognitive impairment and dementia? A systematic review and meta-analysis of longitudinal studies', *Ageing Research Reviews* 52, 7–16.

B. R. Levy et al. (2002). 'Longevity increased by positive self-perceptions of aging', *Journal of Personality and Social Psychology* 83 (2), 261–70.

J. T. Kraiss et al. (2020). 'The relationship between emotion regulation and well-being in patients with mental disorders: a meta-analysis', *Comprehensive Psychiatry* 102, art. no. 152189.

E. J. Boothby, G. Cooney, G. M. Sandstrom and M. S. Clark (2018). 'The liking gap in conversations: do people like us more than we think?', *Psychological Science* 29 (11), 1742–56.

Michael Babula (2013). *Motivation, Altruism, Personality and Social Psychology: the coming age of altruism.* New York: Springer.

V. Klucharev et al. (2009). 'Reinforcement learning signal predicts social conformity', *Neuron* 61 (1), 140–51.

L. Rochat et al. (2019). 'The psychology of "swiping": a cluster analysis of the mobile dating app Tinder', *Journal of Behavioral Addictions* 8 (4), 804–13.

Chicago University (2019). 'UChicago professor developing pill for loneliness', *Chicago Maroon News*, 16 Feb.

R. S. Weiss (1973). *Loneliness: the experience of emotional and social isolation.* Cambridge, MA: MIT Press.

7: SEX ON THE BRAIN

Konstantinos Kapparis (2015). 'Hippocrates, Aristophanes and sex-crazed women', *Ageless Arts: the Journal of the Southern Association for the History of Medicine and Science* 1, 47–57.

Alfred Kinsey et al. (1948). *Sexual Behavior in the Human Male.* Philadelphia: W. B. Saunders.

Alfred Kinsey et al. (1953). *Sexual Behavior in the Human Female.* Philadelphia: W. B. Saunders.

W. H. Masters and V. E. Johnson (1966). *Human Sexual Response.* Toronto and New York: Bantam.

Debby Herbenick et al. (2018). 'Women's experiences with genital touching, sexual pleasure, and orgasm: results from a US probability sample of women ages 18 to 94', *Journal of Sex & Marital Therapy* 44 (2), 201–12.

Nigel Field MD et al. (2013). 'Associations between health and sexual lifestyles in Britain: findings from the third National Survey of Sexual Attitudes and Lifestyles (Natsal-3)', *Lancet* 382 (9907), 1830–44.

Susan E. Trompeter, Ricki Bettencourt and Elizabeth Barrett-Connor (2012). 'Sexual activity and satisfaction in healthy community-dwelling older women', *American Journal of Medicine* 125 (1), P37–43, E1.

Public Health England (2018). *What do women say?* London: Public Health England.

Josie Tetley (2018). 'Let's talk about sex – what do older men and women say about their sexual relations and sexual activities? A qualitative analysis of ELSA Wave 6 data', *Ageing and Society* 38 (3), 497–521.

Markus Parzeller, Roman Bux and Christoph Raschka (2006). 'Sudden cardiovascular death associated with sexual activity: a forensic autopsy study (1972–2004)', *Forensic Science Medicine and Pathology* 2 (2), 109–14.

P. Elwood, J. Galante, J. Pickering, S. Palmer, A. Bayer, Y. Ben-Shlomo et al. (2013). 'Healthy lifestyles reduce the incidence of chronic diseases and dementia: evidence from the Caerphilly cohort study', *PLoS ONE* 8 (12), e81877.

G. Persson (1981). 'Five-year mortality in a 70-year-old urban population in relation to psychiatric diagnosis, personality, sexuality and early parental death', *Acta Psychiatrica Scandinavica* 64, 244.

Julier Frappier et al. (2013). 'Energy expenditure during sexual activity in young healthy couples', *PLoS ONE* 8 (10), e79342.

Helle Gerbild et al. (2018). 'Physical activity to improve erectile function: a systematic review of intervention studies', *Sexual Medicine* 6 (2), 75–89.

Tomás Cabeza de Baca et al. (2017). 'Sexual intimacy in couples is associated with longer telomere length', *Psychoneuroendocrinology* 81, July, 46–51.

B. Whipple and B. R. Komisaruk (1985). 'Elevation of pain threshold by vaginal stimulation in women', *Pain* 21 (4), 357–67.

Vicky Wang et al. (2015). 'Sexual health and function in later life: a population-based study of 606 older adults with a partner', *American Journal of Geriatric Psychiatry* 23 (3), 227–33.

Jennifer R. Rider et al. (2016). 'Ejaculation frequency and risk of prostate cancer: updated results with an additional decade of follow-up', *European Urology* 70 (6), 974–82.

G. G. Giles et al. (2003). 'Sexual factors and prostate cancer', *BJU International* 92 (3), 211–16.

W. Penfield and T. Rasmussen (1950). *The Cerebral Cortex of Man*. New York: Macmillan.

F. L. McNaughton (1977). 'Wilder Penfield: his legacy to neurology. Impact on medical neurology', *Canadian Medical Association Journal* 116 (12), 1370.

Paula M. di Noto et al. (2013). 'The hermunculus: what is known about the representation of the female body in the brain?', *Cerebral Cortex* 23 (5), 1005–13.

Beverly Whipple and Barry R. Komisaruk (1997). 'Sexuality and women with complete spinal cord injury', *Spinal Cord*, 35 (3), 136–8.

Benedetta Leuner, Erica R. Glasper and Elizabeth Gould (2010). 'Sexual experience promotes adult neurogenesis in the hippocampus despite an initial elevation in stress hormones', *PLoS ONE* 5 (7), e11597.

Mark D. Spritzer et al. (2016). 'Sexual interactions with unfamiliar females reduce hippocampal neurogenesis among adult male rats', *Neuroscience* 318, 24 March, 143–56.

Mark S. Allen (2018). 'Sexual activity and cognitive decline in older adults', *Archives of Sexual Behavior*, 47, 1711–19.

Hayley Wright, Rebecca A. Jenks and Nele Demeyere (2019). 'Frequent sexual activity predicts specific cognitive abilities in older adults', *Journals of Gerontology: Series B* 74 (1), 47–51.

Larah Maunder, Dorothée Schoemaker and Jens C. Pruessner (2017). 'Frequency of penile-vaginal intercourse is associated with verbal recognition performance in adult women', *Archives of Sexual Behavior* 46 (2), 441–53.

Olivier Beauchet (2006). 'Testosterone and cognitive function: current clinical evidence of a relationship', *European Journal of Endocrinology* 155 (6), 773–81.

Jacqueline Compton, Therese van Amelsvoort and Declan Murphy (2001). 'HRT and its effect on normal ageing of the brain and dementia', *British Journal of Clinical Pharmacology* 52 (6), 647–53.

8: MIND GAMES

Frederick J. Zimmerman, A. Christakis and Andrew N. Meltzoff (2007). 'Associations between media viewing and language development in children under age 2 years', *Journal of Pediatrics* 151 (4), 364–8.

Jaylyn Waddell and Tracey J. Shors (2008). 'Neurogenesis, learning and associative strength', *European Journal of Neuroscience* 27 (11), 3020–8.

E. A. Maguire, K. Woollett and H. J. Spiers (2006). 'London taxi drivers and bus drivers: a structural MRI and neuropsychological analysis', *Hippocampus* 16 (12), 1091–1101.

Gerd Kempermann et al. (2018). 'Human adult neurogenesis: evidence and remaining questions', *Cell Stem Cell* 23 (1), 25–30.

K. I. Erickson, R. S. Prakash, M. W. Voss et al. (2009). 'Aerobic fitness is associated with hippocampal volume in elderly humans', *Hippocampus* 19 (10), 1030–9.

Alison Abbott (2019). 'First hint that body's "biological age" can be reversed', *Nature* 573 (173).

P. M. Wayne, J. N. Walsh, R. E. Taylor-Piliae et al. (2014). 'Effect of tai chi on cognitive performance in older adults: systematic review and meta-analysis', *Journal of the American Geriatrics Society* 62 (1), 25–39.

Johan Mårtensson et al. (2012). 'Growth of language-related brain areas after foreign language learning', *NeuroImage* 63 (1), 240.

P. K. Kuhl et al. (2016). 'Neuroimaging of the bilingual brain: structural brain correlates of listening and speaking in a second language', *Brain and Language* 162, 1–9.

O. A. Olulade et al. (2016). 'Neuroanatomical evidence in support of the bilingual advantage theory', *Cerebral Cortex* 26 (7), 3196–3204.

S. Alladi, T. H. Bak, V. Duggirala et al. (2013). 'Bilingualism delays age at onset of dementia, independent of education and immigration status', *Neurology* 81 (22), 1938–44.

D. Perani, M. Farsad, T. Ballarini, F. Lubian et al. (2017). 'The impact of bilingualism on brain reserve and metabolic connectivity in Alzheimer's dementia', *Proceedings of the National Academy of Sciences of the United States of America* 114 (7), 1690–5.

T. H. Bak, J. J. Nissan, M. M. Allerhand and I. J. Deary (2014). 'Does bilingualism influence cognitive aging?' *Annals of Neurology* 75, 959–63.

K. D. Lakes et al. (2016). 'Dancer perceptions of the cognitive, social, emotional, and physical benefits of modern styles of partnered dancing', *Complementary Therapies in Medicine* 26, 117–22.

S. Edwards (2016). 'Strength in movement', *Harvard Gazette*, 5 Jan.

A. Z. Burzynska, Y. Jiao et al. (2017). 'White matter integrity declined over 6-months, but dance intervention improved integrity of the fornix of older adults', *Frontiers in Aging Neuroscience*, 9, 59.

Joe Verghese et al. (2003). 'Leisure activities and the risk of dementia in the elderly', *New England Journal of Medicine* 348 (25), 2508–16.

Rehfeld, K. et al. (2018). 'Dance training is superior to repetitive physical exercise in inducing brain plasticity in the elderly', *PLoS ONE* 13 (7), e0196636, https://journals.plos.org/plosone/article?id=10.1371/journal.pone.0196636.

Global Council on Brain Health (2017). *Engage Your Brain: GCBH recommendations on cognitively stimulating activities.* Washington DC: AARP.

Y. Stern (2012). 'Cognitive reserve in ageing and Alzheimer's disease', *Lancet Neurology* 11 (11), 1006–12.

G. M. Whipple (1910). 'The effect of practice upon the range of visual attention and of visual apprehension', *Journal of Educational Psychology* 1 (5), 249–62.

Helen Brooker, Keith A. Wesnes, Clive Ballard et al. (2019). 'The relationship between the frequency of number-puzzle use and baseline cognitive function in a large online sample of adults aged 50 and over', *International Journal of Geriatric Psychiatry* 34 (7), 932–40.

P. Fissler et al. (2017). 'Jigsaw puzzles as cognitive enrichment (PACE): the effect of solving jigsaw puzzles on global visuospatial cognition in adults 50 years of age and older: study protocol for a randomized controlled trial', *Trials* 18 (1), 415.

Arthur R. Jensen (1969). 'How much can we boost IQ and scholastic achievement?', *Harvard Educational Review* 39 (1), 1–123.

Susanne M. Jaeggi, Martin Buschkuehl, John Jonides and Walter J. Perrig (2008). 'Improving fluid intelligence with training on working memory', *Proceedings of the National Academy of Sciences of the United States of America* 105 (19), 6829–33.

Monica Melby-Lervåg, Thomas S. Redick and Charles Hulme (2016). 'Working memory training does not improve performance on measures of intelligence or other measures of "far transfer": evidence from a meta-analytic review', *Perspectives on Psychological Science* 11 (4), 512–34.

T. D. Brilliant, R. Nouchi and R. Kawashima (2019). 'Does video gaming have impacts on the brain? Evidence from a systematic review', *Brain Sciences* 9 (10), 251.

Kyle E. Mathewson et al. (2012). 'Different slopes for different folks: alpha and delta EEG power predict subsequent video game learning rate and improvements in cognitive control tasks', *Psychophysiology* 49 (12), 1558–70.

Aviv M. Weinstein (2017). 'An update overview on brain imaging studies of internet gaming disorder', *Frontiers in Psychiatry* 8, art. no. 185.

D. J. Simons et al. (2016). 'Do "brain-training" programs work?', *Psychological Science in the Public Interest* 17 (3), 103–86.

Monica Melby-Lervåg and Charles Hulme (2012). 'Is working memory training effective? A meta-analytic review', *Developmental Psychology* 49 (2), doi: 10.1037/a0028228.

Adrian M. Owen et al. (2010). 'Putting brain training to the test', *Nature* 49 (12), 1558–70.

9: TO SLEEP OR NOT TO SLEEP

Matthew Walker (2017). *Why We Sleep*. London: Penguin.

R. Legendre and H. Piéron (1908). 'Distribution des altérations cellulaires du système nerveux dans l'insomnie expérimentale', *Comptes Rendus Hebdomadaires des Séances et Mémoires de la Société de Biologie* 64, 1102–4.

M. de Manaceine (1894). 'Quelques observations expérimentales sur l'influence de l'insomnie absolue', *Archives Italiennes de Biologie* 21, 322–5.

Gandhi Yetish et al. (2015). 'Natural sleep and its seasonal variations in three pre-industrial societies', *Current Biology* 25 (21), 2862–8.

M. K. Scullin and D. L. Bliwise, 'Sleep, cognition, and normal aging: integrating a half century of multidisciplinary research', *Perspectives on Psychological Science* 10 (1), 97–137.

Hans Berger (1940). *Psyche* (Jena: Gustav Fischer).

E. Aserinsky and N. Kleitman (1953). 'Regularly occurring periods of eye motility, and concomitant phenomena during sleep', *Science* 118 (3062), 273–4.

Rogers Commission (1986). *Report of the Presidential Commission on the Space Shuttle Challenger Accident*. Washington DC: US Government Publications.

M. Siffre (1988). 'Rythmes biologiques, sommeil et vigilance en confinement prolongé', in *Proceedings of the Colloquium on Space and Sea*, SEE N 88-26016 19-51, 53–68. Marseille: European Space Agency.

N. Goel, M. Basner, H. Rao and D. F. Dinges (2013). 'Circadian rhythms, sleep deprivation, and human performance', *Progress in Molecular Biology and Translational Science* 119: 155–90.

A. Green, M. Cohen-Zion, A. Haim and Y. Dagan (2017). 'Evening light exposure to computer screens disrupts human sleep, biological rhythms, and attention abilities', *Chronobiology International* 34 (7), 855–65.

J. Barcroft (1932). 'La fixité du milieu intérieur est la condition de la vie libre (Claude Bernard)', *Biological Reviews* 7, 24–8.

I. O. Ebrahim, C. M. Shapiro, A. J. Williams and P. B. Fenwick, 'Alcohol and sleep I: effects on normal sleep', *Alcoholism: Clinical and Experimental Research* 37 (4), 539–49.

U. M. H. Klumpers et al. (2015). 'Neurophysiological effects of sleep deprivation in healthy adults, a pilot study', *PLoS ONE* 10 (1), e0116906.

S. L. Worley (2018). 'The extraordinary importance of sleep: the detrimental effects of inadequate sleep on health and public safety drive an explosion of sleep research', *Pharmacy and Therapeutics* 43 (12), 758–63.

Y. Nir, T. Andrillon, A. Marmelshtein et al. (2017). 'Selective neuronal lapses precede human cognitive lapses following sleep deprivation', *Nature Medicine* 23 (12), 1474–80.

S. D. Womack, J. N. Hook, S. H. Reyna and M. Ramos (2013). 'Sleep loss and risk-taking behavior: a review of the literature', *Behavioral Sleep Medicine* 11 (5), 343–59.

Joseph R. Winer, Bryce A. Mander, Randolph F. Helfrich et al. (2019). 'Sleep as a potential biomarker of tau and β-amyloid burden in the human brain', *Journal of Neuroscience* 39 (32) 6315–24.

Global Council on Brain Health (2016). *The Brain–Sleep Connection: GCBH recommendations on sleep and brain health*. Washington DC: AARP.

L. Li, C. Wu, Y. Gan et al. (2016). 'Insomnia and the risk of depression: a meta-analysis of prospective cohort studies', *BMC Psychiatry* 16 (1), 375.

Shalini Paruthi et al. (2016). 'Consensus statement of the American Academy of Sleep Medicine on the recommended amount of sleep for healthy children: methodology and discussion', *Journal of Clinical Sleep Medicine* 12 (11), 1549–61.

N. F. Watson, M. S. Badr, G. Belenky et al. (2015). 'Recommended amount of sleep for a healthy adult: a joint consensus statement of the American Academy of Sleep Medicine and Sleep Research Society', *Sleep* 38 (6), 843–4.

Manu S. Goyal et al. (2019). 'Persistent metabolic youth in the aging female brain', *Proceedings of the National Academy of Sciences of the United States of America* 116 (8), 3251–5.

Monica P. Mallampalli and Christine L. Carter (2014). 'Exploring sex and gender differences in sleep health: a Society for Women's Health Research report', *Journal of Women's Health* 23 (7), 553–62.

Bryce A. Mander, Joseph R. Winer and Matthew P. Walker (2017). 'Sleep and human aging', *Neuron Review* 94, 19–36.

Pierre Philip et al. (2004). 'Age, performance and sleep deprivation', *Journal of Sleep Research* 13 (2), 105–10.

10: THE FEEL-GOOD FACTOR

J. Helliwell, R. Layard and J. Sachs (2019). *World Happiness Report 2019*. New York: Sustainable Development Solutions Network.

Daniel Kahneman and Angus Deaton (2010). 'High income improves evaluation of life but not emotional well-being', *Proceedings of the National Academy of Sciences of the United States of America* 107 (38), 16489–93.

Ashley V. Whillans et al. (2017). 'Buying time promotes happiness', *Proceedings of the National Academy of Sciences of the United States of America* 114 (32), 8523–7.

M. Luhmann, W. Hofmann, M. Eid and R. E. Lucas (2012). 'Subjective well-being and adaptation to life events: a meta-analysis', *Journal of Personality and Social Psychology* 102 (3), 592–615.

Matthew A. Killingsworth and Daniel T. Gilbert (2010). 'A wandering mind is an unhappy mind', *Science* 330 (6006), 932.

D. A. Schkade and D. Kahneman (1998). 'Does living in California make people happy? A focusing illusion in judgments of life satisfaction', *Psychological Science* 9 (5), 340–6.

H. Selye (1936). 'A syndrome produced by diverse nocuous agents', *Nature* 138, 32.

S. Dilger et al. (2003). 'Brain activation to phobia-related pictures in spider phobic humans: an event-related functional magnetic resonance imaging study', *Neuroscience Letters* 348 (1), 29–32.

Antoine Bechara, Hanna Damasio and Antonio R. Damasio (2000). 'Emotion, decision making and the orbitofrontal cortex', *Cerebral Cortex* 10 (3), 295–307.

C. B. Pert and S. H. Snyder (1973). 'Opiate receptor: demonstration in nervous tissue', *Science* 179, 1011–14.

M. A. Crocq (2007). 'Historical and cultural aspects of man's relationship with addictive drugs', *Dialogues in Clinical Neuroscience* 9 (4), 355–61.

Michael J. Brownstein (1993). 'A brief history of opiates, opioid peptides, and opioid receptors', *Proceedings of the National Academy of Sciences of the United States of America* 90, 5391–3.

J. Olds and P. Milner (1954). 'Positive reinforcement produced by electrical stimulation of septal area and other regions of rat brain', *Journal of Comparative and Physiological Psychology* 47 (6), 419–27.

David J. Llewellyn et al. (2008). 'Cognitive function and psychological well-being: findings from a population-based cohort', *Age and Ageing* 37 (6), 685–9.

Laura Mehegan and Chuck Rainville (2018). *AARP Brain Health and Mental Well-Being Survey*. Washington DC: AARP Research.

Arthur A. Stone, Joseph E. Schwartz, Joan E. Broderick and Angus Deaton (2010). 'A snapshot of the age distribution of psychological well-being in the United States', *Proceedings of the National Academy of Sciences of the United States of America* 107 (22), 9985–90.

A. S. Heller, C. M. van Reekum, S. M. Schaefer et al. (2013). 'Sustained striatal activity predicts eudaimonic well-being and cortisol output', *Psychological Science* 24 (11), 2191–2200.

L. B. Pacheco, J. S. Figueira, M. G. Pereira, L. Oliveira and I. A. David (2020). 'Controlling unpleasant thoughts: adjustments of cognitive control based on previous-trial load in a working memory task', *Frontiers in Human Neuroscience* 13, 469.

Walther Mischel (2014). *The Marshmallow Test: why self-control is the engine for success*. Boston: Little, Brown.

Rachel M. Zachar et al. (2016). 'A SPECT study of cerebral blood flow differences in high and low self-reported anger', American Psychological Association 124th Convention.

Ana Loureiro and Susana Veloso (2017). 'Green exercise, health and well-being', in Ghozlane Fleury-Bahi, Enric Pol and Oscar Navarro, eds, *Handbook of Environmental Psychology and Quality of Life Research*, 149–69. New York: Springer.

D. Mosher (2017). 'Professor Kevin K. Fleming PhD interview: what can happen to your brain and body when you shoot a gun', *Business Insider*, Oct. Berlin: Axel Springer.

L. Mineo (2017). 'Good genes are nice, but joy is better: the Harvard Adult Development Study', *Harvard Gazette*, 11 April.

R. M. Yerkes and J. D. Dodson (1908). 'The relation of strength of stimulus to rapidity of habit-formation', *Journal of Comparative Neurology and Psychology* 18 (5). 459–82.

T. Cartwright, H. Mason, A. Porter et al. (2020). 'Yoga practice in the UK: a cross-sectional survey of motivation, health benefits and behaviours', *BMJ Open* 10, e031848.

D. Krishnakumar, M. R. Hamblin and S. Lakshmanan (2015). 'Meditation and yoga can modulate brain mechanisms that affect behavior and anxiety – a modern scientific perspective', *Ancient Science* 2 (1), 13–19.

B. G. Kalyani et al. (2011). 'Neurohemodynamic correlates of "OM" chanting: a pilot functional magnetic resonance imaging study', *International Journal of Yoga* 4 (1), 3–6.

Robert Provine (2000). 'The science of laughter', *Psychology Today*, Nov.

B. Wild, F. A. Rodden, W. Grodd and W. Ruch (2003). 'Neural correlates of laughter and humour', *Brain* 126, 2121–38.

D. M. Buss (1989). 'Sex differences in human mate preferences: evolutionary hypotheses tested in 37 cultures', *Behavioral and Brain Sciences* 12, 1–49.

Global Council on Brain Health (2020). *Music on Our Minds: the rich potential of music to promote brain health and mental well-being*. Washington DC: AARP. Available at www.GlobalCouncilOnBrainHealth.org.

All-Party Parliamentary Group on Arts, Health and Wellbeing (2017). *Creative Health: the arts for health and wellbeing*, inquiry report, 2nd edn, http://www.artshealthandwellbeing.org.uk/appg-inquiry/.

11: COVID-19 AND THE BRAIN

C-Y. Liu et al. (2018). 'Emerging roles of astrocytes in neuro-vascular unit and the tripartite synapse with emphasis on reactive gliosis in the context of Alzheimer's Disease', *Frontiers in Cellular Neuroscience* 12, 193.

G. A. Rossi, O. Sacco, A. Capizzi and P. Mastromarino (2021). 'Can resveratrol-inhaled formulations be considered potential adjunct treatments for COVID-19?', *Frontiers in Immunology* 12, 670955.

M. A. Erickson and W. A. Banks (2019). 'Age-associated changes in the immune system and blood–brain barrier functions', *International Journal of Molecular Science*, 20, 1632.

W. A. Banks, M. J. Reed, A. F. Logsdon, E. M. Rhea and M. A. Erickson (2021). 'Healthy aging and the blood–brain barrier', *Nature Aging* 1, March 2021, 243–54, https://doi.org/10.1038/s43587-021-00043-5.

F. D'Acquisto and A. Hamilton (2021). 'Cardiovascular and immunological implications of social distancing in the context of COVID-19', *Cardiovascular Research* 116 (10), e129–e131.

D. C. Nieman and L. M. Wentz (2021). 'The compelling link between physical activity and the body's defense system', *Journal of Sport and Health Science*, 8, 201–17.

T. Lipinski, D. Ahmad, N. Serey and H. Jouharab (2020). 'Review of ventilation strategies to reduce the risk of disease transmission in high occupancy buildings', *International Journal of Thermofluids*, 7–8, 100045.

G. B. Stefano et al. (2021). '"Brain fog" and SARS-CoV-2 infection', *Medical Science Monitor*, 27, e930886.

A. Hampshire et al. (2021). 'Cognitive deficits in people who have recovered from COVID-19 relative to controls: an N=84,285 online study', *medRxiv* preprint, https://doi.org/10.1101/2020.10.20.20215863.

W. Trypstee, J. van Cleemput, W. Snippenberg et al. (2020). 'On the whereabouts of SARS-CoV-2 in the human body: a systematic review', *PLoS Pathogens* 16 (10), e1009037, https://biblio.ugent.be/publication/8684487/file/8684492.pdf.

M. A. Erickson, E. M. Rhea, R. C. Knopp and W. A. Banks (2021). 'Interactions of SARS-CoV-2 with the blood–brain barrier', *International Journal of Molecular Science*, 22, 2681.

Kanta Subbarao and Siddhartha Mahanty (2020). 'Respiratory virus infections: understanding COVID-19', *Immunity* 52 (6), 905–9, https//.doi.org/10.1016/j.immuni.2020.05.004.

J. Huang, M. Zheng, X. Tang et al. (2020). 'Potential of SARS-CoV-2 to cause CNS infection: biologic fundamental and clinical experience', *Frontiers in Neurology* 11, 659, doi: 10.3389/fneur.2020.00659.

D. Armocida et al. (2020). 'How SARS-Cov-2 can involve the central nervous system: a systematic analysis of literature of the department of human neurosciences of Sapienza University', *Journal of Clinical Neuroscience* 79, 231–6.

A. G. Letizia, I. Ramos and A. Obla (2020). 'SARS-CoV-2 transmission among marine recruits during quarantine', *New England Journal of Medicine* 383, 2407–16, doi: 10.1056/NEJMoa2029717.

H. Bundgaard et al. (2021). 'Effectiveness of adding a mask recommendation to other public health measures to prevent SARS-CoV-2 infection in Danish mask wearers', *Annals of Internal Medicine* 174, 335–43.

A. Varatharaj, N. Thomas, M. A. Ellul et al. (2020). 'Neurological and neuropsychiatric complications of COVID-19 in 153 patients: a UK-wide surveillance study', *Lancet Psychiatry*, 7, 875–82.

R. W. Paterson, R. L. Brown and L. Benjamin (2020). 'The emerging spectrum of COVID-19 neurology: clinical, radiological and laboratory findings', *Brain*, 143, 3104–20, doi:10.1093/brain/awaa240.

M. Boldrini, P. D. Canoll and R. S. Klein (2021). 'How COVID-19 affects the brain', *JAMA Psychiatry* 78 (6), 682–3.

M. Taquet, J. R. Geddes, M. Husain, S. Luciano and P. J. Harrison (2021). '6-month neurological and psychiatric outcomes in 236 379 survivors of COVID-19: a retrospective cohort study using electronic health records', *Lancet Psychiatry* 8, 416–27.

Anthony N. van den Pol (2009). 'Viral infection leading to brain dysfunction: more prevalent than appreciated?', *Neuron* 64 (1), 17–20, doi:10.1016/j.neuron.2009.09.023.

ACKNOWLEDGEMENTS

There are so many people to thank, I hardly know where to start. But this book would not have been created without the vision, patience and editorial skill of Henry Vines, to whom I owe a debt of gratitude. In preparing the manuscript for publication, the keen eye and attention of copy-editor Gillian Somerscales have proved invaluable. I must also thank Jan Stein, art curator of Emaginate LLC, who plagued Henry night and day with millions of emails but who was an inspirational source of advice for the art portfolio. Then there is the Global Council for Brain Health. They were the inspiration for many of my ideas, and in particular I'd like to thank Sarah Lock, Executive Director, and Marilyn Albert, Chair of the Council, for their unstinting support to me as an adviser to the Council.

Chris Talley, Director of Precision Food Works and my 'go-to expert' on all things nutrition, generously gave his time for an inspiring and uplifting interview, reproduced in these pages, for which I was extremely grateful.

I have two universities to thank. First, Loughborough, who placed their faith in me at an early stage of my career, and particularly Barry Bogin, Emeritus Professor, who graciously reviewed my emerging book. I am equally grateful to Clive Ballard, Dean of the University of Exeter Medical School, who patiently fielded my many unlikely questions over the course of writing the book. Thank you, Clive. Dr Duke Han of the University of Southern California also deserves my thanks for generously devoting his time to reviewing the emerging manuscript and protecting me from making too many errors. I'd also like to thank Dr Bill Haley and Dr Ross Andel of the University of South Florida, Tampa, for their enduring encouragement. And not least of all, I'd like to salute the enterprise and vision of William Claiborne Hancock, founder and CEO of Pegasus Books, for ensuring that readers in the USA, including my son Henry in New Hampshire, have the opportunity to access this volume.

I must also mention my pre-publication readers, too many to name individually but who collectively made a huge contribution to the book. Their many suggestions and comments were all gratefully received.

I'd also like to thank all my family, who put up with my long, unsocial hours of writing and grumpiness but who supported me nonetheless. Finally, I'd like to mention my beautiful but now deceased Rottweilers, Cleo and Amber, whose gentle ways and unconditional friendship kept me sane during the long days of writing.

PICTURE ACKNOWLEDGEMENTS

Every effort has been made to contact copyright holders of material reproduced in this book. We would be pleased to rectify any omissions in subsequent editions should they be drawn to our attention. The author and publisher would like to thank the following organizations and individuals for permission to reproduce copyright material.

INTRODUCTION

p. vi, 'Dorothy gazed thoughtfully at the scarecrow': illustration by William Wallace Denslow, from L. Frank Baum, *The Wonderful Wizard of Oz* (1900).

CHAPTER 1

p. 2, 'Phineas Gage holding the tamping iron that injured him': photographer unknown (*c*.1860), courtesy of the Gage family of Texas photo collection.

p. 10, 'Meet your brain': line engraving by J. Tinney (1743), after A. Vesalius (1543), courtesy of The Wellcome Collection.

p. 11, 'The four humours': from *Book of Alchemy* by Thurn-Heisser, Leipzig, Germany (1574).

p. 15, figure 1.1, 'The human brain in situ': adapted from Dr Ananya Mandal, *Human Brain Structure*, news-medical.net.

p. 16, figure 1.2, 'The limbic system': adapted from waitbutwhy.com.

CHAPTER 2

p. 35, 'The *Discobolus*': a Roman bronze reproduction of Myron's *Discobolus* (second century AD), Glyptothek, Munich.

p. 39, 'An active lifestyle': photographed by Mr Phelan for Harrie Irving Hancock, *Physical Training for Business Men* (1917), courtesy of the Library of Congress, Washington DC.

p. 44, figure 2.1, 'Energy expenditure at various ages': redrawn from T. M. Manini, 'Energy expenditure and aging', *Ageing Research Reviews* 9 (1), 2010, p. 9.

p. 49, 'Tennis champions, 1920': 'Wimbledon mixed doubles champions Suzanne Lenglen and Gerald Patterson', *Le Miroir des Sports*, 1920.

CHAPTER 3

p. 63, 'Godzilla, the beast with two brains': Godzilla concept art by Yuji Sakai, based on the original 1954 Godzilla suit crafted by Teizo Toshimitsu and his staff, photograph © Japan Godzilla Festival, 2018.

p. 67, 'Surgeon William Beaumont and his patient Alexis St Martin': by Dean Cornwell, 1938; courtesy of the Library of Congress Prints and Photographs Division, Washington DC.

p. 68, 'Wilbur Olin Atwater and his respiration calorimeter chamber': courtesy of Special Collections, US Department of Agriculture National Agricultural Library, Wilbur Olin Atwater Papers.

p. 75, 'Fasting': *Café table with absinthe* by Vincent van Gogh (1887), courtesy of the Van Gogh Museum, Amsterdam Vincent van Gogh Foundation.

CHAPTER 4

p. 84, 'A woman lying down breast-feeding her baby': by Francesco Bartolozzi (1726–1815), courtesy of The Wellcome Collection.

p. 91, 'A casual kiss': *Billiards – A Kiss* by Nathaniel Currier and James Merritt Ives (1874), lithographers. From *Currier & Ives: A Catalogue Raisonné*, compiled by Gale Research, Detroit, MI, c.1983.

p. 98, 'Eating the whole apple': *Eating the Profits* by J. G. Brown (1878).

p. 104, 'A visit to the dentist': *Unfair Advantage* by F.H. (1892), courtesy of The Wellcome Collection.

CHAPTER 5

p. 117, 'Getting your vitamins': American Donut Association poster, 1941, from the Nutrition Division of the US government's War Food Administration.

p. 125, 'Fruit on a table': *Still Life with Fruit and Decanter* by Roger Fenton (1860).

p. 132, *The Bean Eater*: by Annibale Carraci (1584).

p. 143, 'A well-furnished table': *A man, glass in hand, sits at a table with food and drink*, wood engraving, late sixteenth century, courtesy of The Wellcome Collection.

CHAPTER 6

p. 152, *Loneliness*: by Hans Thoma (1880).

p. 162, 'Solitary confinement': colour woodcut by Elke Rehder from Stefan Zweig, *The Royal Game* (1943), photograph © Sverrir Mirdsson.

p. 171, *Slade School of Art Women's Life Class*: by Bertha Newcombe, photograph of magazine reproduction in *The Sketch* (1895).

p. 176, *Hip, Hip Hurrah! Artists' Party at Skagen*: by Peder Severin Krøyer (1888), courtesy of the Gothenburg Museum of Art.

CHAPTER 7

p. 184, 'Greek vase – the art of seduction': ancient Greek terracotta vase, signed by Hieron as potter and attributed to Makron as painter (*c*.490 BC), courtesy of The Metropolitan Museum of Art, New York.

p. 192, 'The lovers': *Les amants* by Pablo Picasso (1904).

p. 202, Femunculus and homunculus: © Improving Research Limited, UK 2021.

p. 203, figure 7.1, 'Sensory locations in the female brain': adapted from Barry R. Komisaruk et al., *Journal of Sexual Medicine* 8 (10), 2011, pp. 2822–30.

p. 203, figure 7.2, 'Sexual activation in the brain': adapted from *Daily Mail*, 7 Nov. 2010.

CHAPTER 8

p. 224, 'The clever cabbie': *Knowledge is Power* by Glen Marquis.

p. 225, figure 8.1, 'Hippocampus volume at different ages': adapted from L. Nyberg, L-G. Nilsson and P. Letmark, 'Det åldrande minnet: nycklar till att bevara hjärnans resurser', *Natur & Kultur* (Stockholm), 2016 (in Swedish).

p. 232, 'Keep dancing': *The Dance of Life* by Edvard Munch (1899–1900), courtesy of The National Museum of Art, Architecture and Design, Norway.

p. 236, *Violin Concerto*: by Edvard Munch (1903), courtesy of the Museum of Prints and Drawings, Berlin State Museums.

CHAPTER 9

p. 253, 'Hans Berger and his early EEG recordings': from Oksana Zayachkivska, *Adolf Beck, Co-Founder of the EEG* (Utrecht, Digitalis/Biblioscope, 2013).

p. 260, figure 9.1, 'Circadian rhythm': adapted from Michael Reid, 'Strategies for crossing time zones'.

p. 275, 'Sleeping soundly': *Mattress* by Gian Lorenzo Bernini, *c*.1620.

CHAPTER 10

p. 285, figure 10.1, 'Kahneman's experiment on memory': adapted from Donald A. Redelmeier and Daniel Kahneman, 'Patients' memories of painful medical treatments: real-time and retrospective evaluations of two minimally invasive procedures', *Pain* 66 (1), 1996, pp. 3–8.

p. 288, 'Eight physiognomies of human passions': etching by Taylor (1788), after Charles Le Brun (1619–90), courtesy of The Wellcome Collection.

p. 296, figure 10.2, 'Well-being and age': adapted from 'A snapshot of the age distribution of psychological well-being in the United States, a 2010 study by Arthur Stone PhD, University of Southern California (USC).

p. 306, figure 10.3, 'Performance, stress and arousal: the Yerkes–Dodson law': adapted from David M. Diamond et al., 'The temporal dynamics model of emotional memory processing: a synthesis on the neurobiological basis of stress-induced amnesia, flashbulb and traumatic memories, and the Yerkes-Dodson Law' *Neural Plasticity*, 2007, article ID 060803.

p. 309, 'Man's best friend': *Portrait of a Man with a Dog* by Bartolomeo Passarotti (1585), photograph © The Italian Art Society (IAS).

p. 310, An early study for *WHAAM!* by Roy Lichtenstein: © Estate of Roy Lichtenstein/DACS/Artimage 2021.

p. 312, 'Two country dancers': *La danse à la campagne* by Pierre-Auguste Renoir (c.1880), courtesy of the Library of Congress Prints and Photographs Division, Washington DC.

CHAPTER 11

p. 320, 'Dr Beak, a plague doctor, 1656': Paul Fürst, 'Der Doctor Schnabel von Rom'. Wikimedia Commons; freely licensed media.

p. 328, 'SARS-CoV-2 virus particles in droplets': *Penn Medicine*, 2 August 2020.

p. 337, 'Brain fog': Posterior 'X-ray image of human head with cloud', 2021. Shutterstock.

p. 338, figure 11.1, 'Main routes of infection for respiratory diseases such as Covid-19': redrawn from T. Lipinski, D. Ahmad, N. Serey, and H. Jouharab, 'Review of ventilation strategies to reduce the risk of disease transmission in high-occupancy buildings', *International Journal of Thermofluids*, Vols 7–8, Nov. 2020, https://doi.org/10.1016/j.ijft.2020.100045.

p. 340, 'Masking the brain': Google photos, 2021.

p. 343, 'Inflammation': PIXOLOGIC Studio, Science Photo Library, Getty Images Plus, 2020.

p. 346, 'Sex and Covid: the lovers' dilemma': Glasgow West End mural, 2021. Street Muralist Rebel Bear.

All graphs and figures © Global Blended Learning Ltd.

INDEX

ABOUT THE AUTHOR

James Goodwin never had aspirations to write a book or become a professor. After leaving university with a first degree in biology, he joined the British Army. He graduated from Sandhurst and then served as an infantry officer in the UK, the United States, Germany and the Near East on intelligence security duties, and post-war in the Falkland Islands. He could hit targets quite well and was chosen to shoot for the Army internationally in military competitions. He managed to win gold, silver and bronze medals in national meetings. As a young officer, he won a prize in the Bertrand Stewart essay competition. On leaving the regular Army, he made the fateful decision to venture back into university, to 'make up for my indifferent academic performance first time around'. He graduated with a master's degree in human biology from Loughborough followed by a PhD in physiology from Professor Sir John Tooke's department at Exeter University Medical School. He now has a chair in both universities. He became the first head of research in the charity Help the Aged and the Chief Scientific Officer in Age UK, and was seconded to the Centre for Better Ageing in London, advising throughout on commissioning research and learning how to translate it into tangible benefits. He was appointed Fellow of the Academy of Social Sciences for his work on research impact in society. As the chair of a WHO advisory group on knowledge transfer and a member of the Ministerial Advisory Group on Dementia Research, he is a respected special adviser to the Global Council on Brain Health in Washington DC. James now works as the Director of Science and Research Impact at the Brain Health Network in London (www.brain.health). When he is not writing books or public speaking, he lives in Devon, happily walking his dogs on Dartmoor, England's last remaining wilderness area and his source of inspiration from the natural world. James enjoys visiting family, friends and colleagues worldwide. Last but not least, he's grateful for life's rich opportunities and appreciates the privilege of sharing his lifelong learning legacy.